ROWAN UNIVERSITY
CAMPBELL LIBRARY
201 MULLICA HILL RD.
GLASSBORO, NJ 08028-1701

GUIDELINES FOR PERFORMING EFFECTIVE PRE-STARTUP SAFETY REVIEWS

This book is one in a series of process safety guideline and concept books published by the Center for Chemical Process Safety (CCPS). Please go to www.wiley.com/go/ccps to see the full list of titles.

GUIDELINES FOR PERFORMING EFFECTIVE PRE-STARTUP SAFETY REVIEWS

Center for Chemical Process Safety

An **AIChE** Industry
Technology Alliance

A JOHN WILEY & SONS, INC., PUBLICATION

Copyright © 2007 by American Institute of Chemical Engineers. All rights reserved.

A Joint Publication of the Center for Chemical Process Safety of the American Institute of Chemical Engineers and John Wiley & Sons, Inc.

Published by John Wiley & Sons, Inc., Hoboken, New Jersey.
Published simultaneously in Canada.

No part of this publication may be reproduced, stored in a retrieval system, or transmitted in any form or by any means, electronic, mechanical, photocopying, recording, scanning, or otherwise, except as permitted under Section 107 or 108 of the 1976 United States Copyright Act, without either the prior written permission of the Publisher, or authorization through payment of the appropriate per-copy fee to the Copyright Clearance Center, Inc., 222 Rosewood Drive, Danvers, MA 01923, (978) 750-8400, fax (978) 750-4470, or on the web at www.copyright.com. Requests to the Publisher for permission should be addressed to the Permissions Department, John Wiley & Sons, Inc., 111 River Street, Hoboken, NJ 07030, (201) 748-6011, fax (201) 748-6008, or online at http://www.wiley.com/go/permission.

Limit of Liability/Disclaimer of Warranty: While the publisher and author have used their best efforts in preparing this book, they make no representations or warranties with respect to the accuracy or completeness of the contents of this book and specifically disclaim any implied warranties of merchantability or fitness for a particular purpose. No warranty may be created or extended by sales representatives or written sales materials. The advice and strategies contained herein may not be suitable for your situation. You should consult with a professional where appropriate. Neither the publisher nor author shall be liable for any loss of profit or any other commercial damages, including but not limited to special, incidental, consequential, or other damages.

For general information on our other products and services or for technical support, please contact our Customer Care Department within the United States at (800) 762-2974, outside the United States at (317) 572-3993 or fax (317) 572-4002.

Wiley also publishes its books in a variety of electronic formats. Some content that appears in print may not be available in electronic format. For information about Wiley products, visit our web site at www.wiley.com.

Library of Congress Cataloging-in-Publication Data is available.

ISBN 978-0-470-13403-0

Printed in the United States of America.

10 9 8 7 6 5 4 3 2 1

It is sincerely hoped that the information presented in this document will lead to an even more impressive safety record for the entire industry; however, neither the American Institute of Chemical Engineers, its consultants, CCPS Technical Steering Committee and Subcommittee members, their employers, their employers officers and directors, nor AntiEntropics, Incorporated and its employees warrant or represent, expressly or by implication, the correctness or accuracy of the content of the information presented in this document. As between (1) American Institute of Chemical Engineers, its consultants, CCPS Technical Steering Committee and Subcommittee members, their employers, their employers officers and directors, and AntiEntropics, Inc., and its employees, and (2) the user of this document, the user accepts any legal liability or responsibility whatsoever for the consequence of its use or misuse.

CONTENTS

List of Tables .... *xiii*
List of Figures .... *xv*
Items on the CD Accompanying This Book .... *xvii*
Acronyms and abbreviations .... *xix*
Glossary .... *xxi*
Acknowledgements .... *xxiii*
Preface .... *xxv*

1	INTRODUCTION		1
	1.1	What are the Benefits of Performing Pre-startup Safety Reviews?	2
	1.2	How PSSR Relates to Other Process Safety Elements	4
	1.3	An Overview of the Risk-based Approach to PSSR	6
	1.4	What is the Scope of a PSSR? Process Safety, Environmental, Quality and Personnel Safety Considerations	6
	1.5	This Guideline's Audience	7
	1.6	How to use this Guideline	8
	1.7	References	10

2	WHAT IS A PRE-STARTUP SAFETY REVIEW?		11
	2.1	The Basics of Pre-startup Safety Review	11
	2.1.1	Some Common Steps for Performing PSSR	11
	2.2	What is a Risk-based Approach to PSSR?	16
	2.3	The Role of Training in Pre-startup Safety Review	16
	2.3.1	Training Team Leaders and Members	17
	2.3.2	Training Managers and the Remaining Workforce	18
	2.4	Scheduling Considerations	18
	2.4.1	Capital Projects	18

	2.4.2	Changes to Operating Facilities	19
	2.4.3	Temporary Changes	20
	2.4.4	Restarting a Mothballed Process	20
	2.4.5	Post-turnaround Startup	21
	2.4.6	Routine Maintenance	21
	2.4.7	Startup After Emergency Shutdown	21
2.5		References	22

3 REGULATORY ISSUES — 23

3.1	An Overview of PSSR Industry Guidelines and Regulations	23
3.2	Best Practices for PSSR	28
3.3	Environmental Considerations	29
3.4	General Safety, Security, and Occupational Health Considerations	29
3.5	References	30

4 A RISK-BASED APPROACH TO PRE-STARTUP SAFETY REVIEW — 31

4.1		Using Risk Analysis Techniques to Select the Level of Detail for a PSSR	31
	4.1.1	A Case of Complexity Versus Simplicity	32
	4.1.2	The Term Complexity Includes Novelty	35
	4.1.3	The Effect of Complexity on PSSR Team Size and Expertise	36
	4.1.4	The Effect on the Level and Scope of the Review	38
4.2		A Decision Guideline for Designing a PSSR	38
	4.2.1	A Definition of Risk-based PSSR – A Qualitative Approach	38
	4.2.2	An Example Algorithm	39
4.3		Typical Considerations for all Pre-startup Safety Reviews	41
	4.3.1	Hardware and Software: Equipment, Instrumentation, and Process Control	41
	4.3.2	Documentation: Process Safety Information, Procedures, and Maintenance Management System Data	41
	4.3.3	Training: Quality and Verification of Completeness	42

CONTENTS   ix

4.3.4	Special Items: Specific Safety, Health, and Environmental Issues	43
4.4	An Example Risk-based Questionnaire	43
4.5	Two Examples of Using a Risk-based Approach to PSSR Design	50
4.5.1	A Simple PSSR	51
4.5.2	A More Complex PSSR	53
4.6	References	61

5 THE PRE-STARTUP SAFETY REVIEW WORK PROCESS   63

5.1	Defining the PSSR System	63
5.1.1	Double Checking Management of Change	64
5.1.2	Who Is Responsible for Driving the System?	65
5.2	PSSR Sub-elements	66
5.2.1	Construction and equipment meet the designed specifications.	66
5.2.2	Safety, operating, maintenance and emergency procedures are in place and adequate.	66
5.2.3	A PHA has been performed for new facilities.	66
5.2.4	Training of each employee involved in the process is complete.	66
5.2.5	General requirements	67
5.3	Designing and Implementing an Initial PSSR Program	67
5.3.1	Defining a Policy on PSSR	67
5.3.2	Defining the PSSR Team	68
5.3.3	Designing the Specific PSSR	68
5.3.4	Training the Workforce on the PSSR Program	69
5.3.5	An Example PSSR Program	69
5.4	Preparing to Perform a Pre-statup Safety Review	75
5.4.1	Gather the Documentation	75
5.4.2	Schedule Meetings as Needed	75
5.4.3	Verify the Trigger Event Related Work Is Complete	76
5.4.4	Identify and Track the Process Hazard Analysis Action Items	76
5.5	Follow Pre-startup Safety Review Action Items	77
5.5.1	Which Items Are Critical for Safe Operation?	78
5.5.2	Consider Past PSSR PSM Compliance Audit Findings	78
5.6	Approve the Pre-startup Safety Review Report	78
5.6.1	Reference the Documentation: Electronic or Hardcopy	79

	5.6.2 PSSR Team Approval	79
	5.6.3 Management Approval	79
	5.7 References	79

6 METHODOLOGIES FOR COMPILING AND USING A PSSR CHECKLIST — 80

6.1	Building Your Facility's Database of Questions	80
6.1.1	Beware of Shortcuts	80
6.1.2	Considerations for Different Industries	81
6.2	Various Approaches: Electronic versus Hardcopy	81
6.2.1	Using your Existing Facility Action Item Tracking System	81
6.2.2	Basic Electronic PSSR Checklist Tools	82
6.2.3	Electronic Change Management Systems with PSSR Tools	82
6.3	An Example Electronic Checklist	83
6.3.1	Collapse the Checklist for Simple PSSR	83
6.3.2	Expand the Checklist for Complex PSSR	85

7 CONTINUOUS IMPROVEMENT — 90

7.1	Diagnosing PSSR System Issues	90
7.2	Training and Communication	91
7.3	Examine Excesses as well as Deficiencies	92
7.4	Why Refine, Improve, Upgrade, or Redesign?	92
7.4.1	Workforce Reductions	93
7.4.2	Company Restructuring	93
7.4.3	Acquisitions, Mergers, and Divestiture	93
7.4.4	Regulatory Changes	93
7.4.5	Changes in Process Risk	93
7.5	Upgrading the System	94
7.6	Example PSSR Performance and Efficiency Metrics	95
7.6.1	PSSR Performance Indicators	95
7.6.2	PSSR Efficiency Indicators	96
7.7	Audit Frequency	96
7.8	Qualification Considerations for PSSR Auditors	97
7.9	Sample PSSR Audit Protocols	98
7.10	Addressing Audit Results	102

	7.11	Summary	103
	7.12	References	103

APPENDIX A – PSSR CHECKLIST EXAMPLES   105

APPENDIX B – INDUSTRY REFERENCES   153

APPENDIX C – REGULATORY REFERENCES   157

INDEX   159

LIST OF TABLES

TABLE 1-1	How PSSR Typically Interfaces with Other PSM Elements	5
TABLE 3-1	Global Examples of Pre-startup Safety Review Related Documents	25
TABLE 4-1	Typical Risk-based Decision Making Steps in a PSM PSSR Procedure	34
TABLE 4-2	Example short form for lower risk / simple PSSR	44
TABLE 4-3	Example long form for higher risk / complex PSSR	45
TABLE 4-4	Example completed short form PSSR	53
TABLE 4-5	Example long form PSSR in progress	56
TABLE 5-1	An example integrated PSM/RMP compliance plan and management system	64
TABLE 5-2	Example pre-startup safety review administrative procedure	70
TABLE 6-1	Example of collapsing an electronic PSSR checklist	83
TABLE 6-2	Example expansion of an electronic PSSR checklist in progress	85
TABLE 7-1	Typical PSSR issues from formal or informal audits	91
TABLE 7-2	Example guideline wording for a PSSR element audit protocol	100
TABLE 7-3	Excerpted PSSR language from OSHA CPL 2-2.45A CH-1	101

LIST OF FIGURES

FIGURE 2-1 Basic PSSR work flow chart. 15
FIGURE 4-1 An example risk matrix chart 33

ITEMS ON THE CD ACCOMPANYING THIS BOOK

The text of the book

Example PSSR checklists

An example PSSR management system procedure

An Excel spreadsheet of basic PSSR checklist items in an expandable format

ACRONYMS AND ABBREVIATIONS

ACC	American Chemistry Council
AIChE	American Institute of Chemical Engineers
API	American Petroleum Institute
ASME	American Society of Mechanical Engineers
CCPS	Center for Chemical Process Safety
CFR	Code of Federal Regulations
CPI	chemical process industries
DOT	Department of Transportation
EPA	Environmental Protection Agency
ESD	emergency shutdown
HAZMAT	hazardous material
HAZOP	hazard and operability study
ISO	International Organization for Standardization
MOC	management of change
MSDS	material safety data sheet(s)
NDT	nondestructive testing
NEC	National Electric Code
NFPA	National Fire Protection Association
OSHA	Occupational Safety and Health Administration
P&ID	piping and instrumentation diagram

PHA	process hazard analysis
PM	preventive maintenance
PPE	personal protective equipment
PSM	process safety management
PSSR	pre-startup safety review
PSV	pressure safety valve
QA	quality assurance
RAGAGEP	recognized and generally accepted good engineering practice
RMP	risk management program
SCBA	self-contained breathing apparatus
SIS	safety instrumented system
SME	subject matter expert
UL	Underwriters Laboratories Inc.

GLOSSARY

Gantt Chart   A manner of depicting multiple, time-based project activities (usually on a bar chart with a horizontal time scale).

Hazard and operability (HAZOP) study   A systematic method in which process hazards and potential operating problems are identified using a series of guide words to investigate process deviations.

Management of change (MOC)   A management system for ensuring that changes to processes are properly analyzed (for example, for potential adverse impacts), documented, and communicated to affected personnel.

Mechanical integrity (MI)   A management system for ensuring the ongoing durability and functionality of equipment.

Nondestructive testing/examination (NDT/NDE)   Evaluation of an equipment item with the intention of measuring an equipment parameter without damaging or destroying the equipment item.

Performance measure   A metric used to monitor or evaluate the operation of a program activity or management system.

Pre-startup safety review (PSSR)   A final check, initiated by a trigger event, prior to the use or reuse of a new or changed aspect of a process. It is also the term for the OSHA PSM and EPA RMP element that defines a management system for ensuring that new or modified processes are ready for startup.

Process hazard analysis (PHA)   A systematic evaluation of process hazards with the purpose of ensuring that sufficient safeguards are in place to manage the inherent risks.

Process safety information (PSI)   A compilation of chemical hazard, technology, and equipment documentation needed to manage process safety.

Quality assurance (QA)   Activities to ensure that equipment is designed appropriately and to ensure that the design intent is not compromised throughout the equipment's entire life cycle.

Recommended and generally accepted good engineering practice (RAGAGEP)   Document that provides guidance on engineering, operating, or maintenance activities based on an established code, standard, published technical report, or recommended practice (or a document of a similar name).

Replacement in kind   A replacement that satisfies the design specifications.

Risk A measure of potential loss (for example, human injury, environmental insult, economic penalty) in terms of the magnitude of the loss and the likelihood that the loss will occur.

Risk analysis The development of a qualitative and/or quantitative estimate of risk-based on engineering evaluation and mathematical techniques (quantitative only) for combining estimates of event consequences and frequencies.

Trigger event Any change being made to an existing process, or any new facility being added to a process or facility, or any other activity a facility designates as needing a pre-startup safety review. One example of a non-change related trigger event is performing a PSSR before restart after an emergency shutdown.

Verification activity A test, field observation, or other activity used to ensure that personnel have acquired necessary skills and knowledge following training.

ACKNOWLEDGEMENTS

The American Institute of Chemical Engineers (AIChE) wishes to thank the Center for Chemical Process Safety (CCPS) and those involved in its operation, including its many sponsors whose funding made this project possible, and the members of the Technical Steering Committee, who conceived of and supported this guideline project. The members of the Pre-startup Safety Review subcommittee who worked with AntiEntropics Inc. to produce this text deserve special recognition for their dedicated efforts, technical contributions, and overall enthusiasm for creating a useful addition to the process safety guideline series. CCPS also wishes to thank the subcommittee members' respective companies for supporting their involvement in this project.

The chairman of the Pre-startup Safety Review subcommittee was Perry Morse of E.I. DuPont de Nemours and Company. The CCPS staff liaison was Dan Sliva. The members of the CCPS pre-startup safety review guideline subcommittee were:

Larry Bowler
GE Advanced Materials

Don Connolley
BP

Susan Cowher
ISP Technologies, Inc.

Jonathan Gast
Wyeth Pharmaceuticals

David M. Hawkins, CSP
Intel Corporation

John Herber
3M Company

Lisa Morrison
PPG Industries, Inc.

Steve Marwitz
Formosa Plastics

Glen Peters
Air Products

Cedric Pereira
BP

Michael Moriarty
Akzo Nobel Chemicals Inc.

Michael Rogers
Syncrude Canada Ltd.

James Slaugh
Basell North America Inc.

Robert J. Stankovich
Eli Lilly and Company

Angela Summers
SIS Tech Solutions LP

AntiEntropics, Inc. of New Market, Maryland, was the contractor for this project. Robert J. Walter was the principal author of the text and would like to recognize Sandra A. Baker and James M. Godwin for their expert editorial support.

CCPS also gratefully acknowledges the comments submitted by the following peer reviewers:

John Alderman RRS Engineering	Victor J. Maggioli Feltronics Corporation
Helmut Bezecny The Dow Chemical Company (ret.)	Pike Prescott Sun Chemicals Inc
William L. Bobinger E.I. DuPont de Nemours	William Ralph BP Products North America
Frederic Gil BP Refining and Technology	Sandra K. Schmitzer GE Plastics
Dennis Heavin Eli Lilly and Company	Casey R. Stephenson Eli Lilly and Company
Kevin Klein Solutia Inc	Angela Summers SIS Tech Solutions LP
Peter N. Lodal Eastman Chemical Company	John C. Wincek Croda USA
Randolph Matsushima Suncor Energy U.S.A., Inc.	

Their insights, comments, and suggestions helped ensure a balanced perspective for this guideline.

PREFACE

The American Institute of Chemical Engineers (AIChE) has been closely involved with process safety and loss control issues in the chemical and allied industries for more than four decades. Through its strong ties with process designers, constructors, operators, safety professionals, and members of academia, AIChE has enhanced communications and fostered continuous improvement of the industry's high safety standards. AIChE publications and symposia have become information resources for those devoted to process safety and environmental protection.

AIChE created the Center for Chemical Process Safety (CCPS) in 1985 after the chemical disasters in Mexico City, Mexico, and Bhopal, India. The CCPS is chartered to develop and disseminate technical information for use in the prevention of major chemical accidents. The center is supported by more than 80 chemical process industries (CPI) sponsors who provide the necessary funding and professional guidance to its technical committees. The major product of CCPS activities has been a series of guidelines to assist those implementing various elements of a process safety and risk management system. This book is part of that series.

Pre-startup safety review (PSSR) is a fundamental element of successful process safety programs. However, facilities continue to be challenged to maintain successful PSSR programs in a way that improves total process safety over time. The CCPS Technical Steering Committee initiated the creation of these guidelines to assist facilities in meeting this challenge. This book contains approaches for designing, developing, implementing, and continually improving a pre-startup safety review system. The CD accompanying this book contains resource materials and support information.

1
INTRODUCTION

The term pre-startup safety review (PSSR), in its simplest definition, means a final check prior to initiating the use of process equipment. When the term is used as a part of the overall process safety management program at a facility, it implies a management system within that program for ensuring that new or modified processes are ready for startup. This is accomplished by verifying that equipment is installed in a manner consistent with the design intent and that process safety management systems are in place. As each change, the associated risk, and the process in question may be unique, applying PSSR in a systematic way to your work processes simply represents a good business practice. Performing an effective pre-startup safety review is analogous to checking your math after performing a calculation or, for a more vivid analogy, checking your parachute before a jump.

In this book we examine the application of the practice of PSSR to the physical plant hardware, software, engineering, and management activities, and documentation associated with operating chemical processes. Although we focus on the chemical process industries (CPI) and specifically the associated process safety aspects, the concept of PSSR and its benefits apply to almost any human endeavor, especially in the manufacturing realm.

Since 1992, a major incentive for the chemical process industries (CPI) in the United States to make pre-startup safety review a part of day-to-day business practices has been the Occupational Safety and Health Administration's (OSHA's) process safety management (PSM) regulation (29 *Code of Federal Regulations* [CFR] 1910.119) (Reference 1-1). This regulation was followed by the Environmental Protection Agency's (EPA's) risk management program (RMP) rule (40 CFR 68) (Reference 1-2). These regulations are performance based and apply to facilities processing certain chemicals when present at or above specific threshold quantities. The term performance based means that each facility that falls under the regulation needs to meet certain minimum requirements, but how they meet those requirements is not prescriptive. A facility can build the PSSR program that best fits its risk levels, organizational culture, and resources.

However, other countries and industry organizations have recognized the importance of PSSR to process safety and have published similar rules or guidelines for using it. We will keep the general aspect of the concept of this "final check" in mind, address specific U.S. regulatory needs, but at the same time, give some specific examples of how global chemical processing companies apply pre-startup safety review to their operations to both comply with the applicable laws and enhance their manufacturing performance.

Whether implementing process safety management as a requirement or as a good practice, pre-startup safety review is essential to keeping the system alive and functioning properly to protect a facility against risk.

1.1 WHAT ARE THE BENEFITS OF PERFORMING PRE-STARTUP SAFETY REVIEWS?

There are many benefits to be gained from performing pre-startup safety review for your processes. A simple list includes:
- The change is more likely to operate as intended.
- The construction, maintenance, or programming work performed to build, install, or program the process change meets the design specifications originally intended.
- Pre-startup activities have been completed and post-startup activities are scheduled and tracked to help ensure that equipment is designed, fabricated, procured, installed, operated, and maintained in a manner appropriate for its intended application.
- New chemicals or materials used in the process are understood in regard to safety, health, environmental and material performance issues.
- Personnel assigned to inspect, test, maintain, procure, fabricate, install, or commission process equipment are appropriately trained and have access to current and up-to-date procedures and process safety information.
- In the event of an incident, a strong pre-startup safety review program documents corporate operational discipline and social responsibility.
- The safety systems are confirmed to be operating as designed.
- Engineering calculations and assumptions used for design and installation match recognized and generally accepted good engineering practices (RAGAGEP) which describe applicable codes and standards.
- Regulatory requirements for managing changes are met.
- Quality management system requirements for your company have been addressed.

1. INTRODUCTION

- PSSR provides an opportunity for turnover of ownership from engineering or project managers to operations personnel

This book provides advice for developing a PSSR program that will assist organizations in achieving these benefits. Facilities should consider evaluating how they are doing in regard to reaping the benefits above. Chapter 7 – *Continuous Improvement* addresses this issue.

On a broader scale, effective PSSR supports any mature, well-designed PSM program by keeping the total program robust and vibrant in the face of change. The CCPS booklet *The Business Case for Process Safety* summarizes the benefits of process safety – and these same benefits are supported by performing effective pre-startup safety reviews:

"...methodically implementing process safety provides four benefits essential to any healthy business. Two of these benefits are qualitative and as a result are somewhat subjective. You can see them in the way the public, your shareholders, government bodies, and your customers relate to your company. The two remaining benefits are quantitative. These have measurable impact in terms of your bottom line and company performance. All four benefits, when realized together by adhering to a sound process safety system, combine to support the profitability, safety performance, quality, and environmental responsibility of your business.

- ***Corporate Responsibility*** *– Process safety is the embodiment of corporate responsibility and accountability. It helps your company display these characteristics through its actions. The heart of process safety lies in consistently planning to do the right things, then doing them right – consistently. Corporate responsibility leads to the second benefit...*

- ***Business Flexibility*** *– Corporate responsibility as demonstrated in your process safety management program leads to a greater range of business flexibility. When you openly display responsibility through implementing an effective process safety program, your company can achieve greater freedom and self-determination.*

- ***Risk Reduction*** *– Process safety provides unparalleled loss avoidance capability. A healthy process safety program significantly reduces the risk of catastrophic events and helps prevent the likelihood of human injury, environmental damage, and associated costs that arise from incidents. Although the essence of process safety focuses on preventing catastrophic incidents, the number of less severe incidents is also reduced.*

- ***Sustained Value*** *– Process safety relates directly to enhanced shareholder value. When properly implemented, it helps ensure reliable processes that can produce high quality products, on time, and at lower cost. This increases shareholder value."*

Pre-startup safety review provides a second level of protection to ensure operational readiness, which will drive continuous improvement in your process safety management system, and help your organization realize these four benefits.

1.2 HOW PSSR RELATES TO OTHER PROCESS SAFETY ELEMENTS

This guideline assumes the reader is already familiar with the fourteen basic elements of process safety as defined in the OSHA process safety management regulation and the EPA risk management program rule. These are:

1. Employee Participation
2. Process Safety Information
3. Process Hazard Analysis
4. Management of Change
5. Operating Procedures
6. Mechanical Integrity
7. Emergency Planning & Response
8. Training
9. Contractors
10. Hot Work Permit
11. Compliance Audits
12. Pre-Startup Safety Review
13. Incident Investigation
14. Trade Secrets

A well-designed PSSR program will fit within a facility's existing process safety and risk management program as well as any other performance enhancement effort (for example, six sigma, total quality, environmental management, or profitability initiatives). Personnel charged with developing, implementing and upgrading the PSSR program can better achieve a higher level of overall process safety performance when they know how pre-startup safety review affects or is affected by the other elements of process safety. Table 1-1 illustrates how the other elements of PSM may relate to PSSR.

1. INTRODUCTION

TABLE 1-1
How PSSR Typically Interfaces with Other PSM Elements

PSM Element	Potential Interface
Employee Participation	• Employees from various departments can have input into the PSSR program as developers, team leaders, team members, or interviewees during the reviews. • The PSSR procedure and PSSR checklist documentation provides clear evidence of how your organization encourages employee participation.
Process Safety Information	• PSSR assists in verifying that process safety information (PSI) for equipment, material hazards, and technology is updated in a timely fashion.
Process Hazard Analysis (PHA)	• PSSR assists in verifying any PHA action items required have been or will be addressed.
Operating Procedures	• PSSR provides a second check on whether the operating procedures affected by the change have been written or revised to properly reflect the change.
Operator Training	• PSSR checks to verify any changes to training related to the trigger event have been made and that training on the affected procedures has occurred as needed.
Mechanical Integrity	• PSSR verifies maintenance task procedures are in place and workers have been trained on the tasks and applicable safe work practices. • PSSR verifies equipment has been reviewed for placement in the mechanical integrity program and that it was designed and installed according to codes, standards, and manufacturers' recommendations.
Contractors	• PSSR can identify when certain contract job tasks require special training in response to a change and when contractors need to be trained or informed on aspects of a change.
Hot Work Permit (and other safe work practices)	• PSSR verifies new safe work practices (SWPs) required for the trigger event are in place and designed and implemented for the targeted workers.
Management of Change (MOC)	• PSSR is a check of every MOC-related activity and its documentation; it is a second level of protection to ensure MOC is working to keep workers and the public safer. • The complexity of the PSSR is determined based upon information in the initial MOC request and associated documentation. • PSSR can confirm that a proper management of change effort was performed.
Incident Investigation	• PSSR documentation may provide support to investigation teams. • Investigation recommendations may impact future PSSR activities. • Lessons learned are powerful tools for improvement.
Emergency Planning and Response	• A well-designed PSSR verifies that applicable emergency response plan changes are included in the review and affected workers are trained.
Compliance Audits	• The PSSR program will be audited on a regular basis and those audit results can help improve the PSSR program and a facility's overall PSM performance.
Trade Secrets	• A well-designed PSSR can verify that applicable trade secrets are addressed properly.

1.3 AN OVERVIEW OF THE RISK-BASED APPROACH TO PSSR

What do we mean by a *risk-based* approach to pre-startup safety review? This term indicates that we will use the performance-based aspect of the PSSR regulations and industry guidelines to more efficiently and effectively design each change or trigger event's PSSR activities based upon the likelihood and consequences. The goal is to make the best use of organizational resources based upon the risk attributed to the trigger event for the process.

This guideline offers several examples of qualitative tools to help a PSSR leader and his or her team determine whether a PSSR can be done simply (for example, a visual inspection and documentation of completion using a minimal checklist) or whether it needs to be performed in a more complex fashion due to the hazards and potential risks involved. For example, a major addition to a unit might involve several PSSR team meetings and reviews with some action items to be completed after startup.

Detailed examples for your PSM manager and PSSR team leaders to consider are provided in Chapter 4 – *A Risk-based Approach to Pre-startup Safety Review* and throughout the book.

1.4 WHAT IS THE SCOPE OF A PSSR? PROCESS SAFETY, ENVIRONMENTAL, QUALITY AND PERSONNEL SAFETY CONSIDERATIONS

Our objective is not to set new standards for the chemical processing industry, but to encourage companies and individuals to apply existing standards and the operational discipline necessary to establish their own internal requirements for pre-startup safety review that support business excellence. This guideline provides tools to help companies take a systematic approach to managing their PSSR program and implement applicable portions of Responsible Care®, CCPS' process safety practices, and other industry guidance, while meeting external and internal health, safety, environmental, and quality requirements.

For this guideline's purposes, a pre-startup safety review is considered to be applying a systematic method to confirm that the startup team and process equipment are prepared for startup. Startup may be considered to be the point at which chemicals or energy is introduced into the system. This practice is key to any loss prevention effort in the manufacturing realm.

It is a final check to confirm that a process or facility has been built as designed, all procedures are in place, training is complete, and all action items from the process hazard analysis for the activity have been resolved.

1. INTRODUCTION

A PSSR for U.S. OSHA PSM and EPA RMP compliance is required when any change modifies the process safety information. What is a change to the process safety information? It is any change to any of the following items:

- Information pertaining to the hazards of the regulated substances used or produced by the process
    - A new catalyst or treatment additive
    - A new feedstock, even if inherently safer
- Information pertaining to the technology of the process
    - A new control or safety system
- Information pertaining to the equipment in the process
    - A new type of reactor or process vessel
    - New valves or valve operations

The PSSR for OSHA and EPA compliance should consider whether:

- Construction and equipment meet the design specifications.
- Safety, operating, maintenance and emergency procedures are in place and adequate.
- A PHA has been performed for new facilities or that the site management of change process has been followed.
- Training of each employee involved in the operating process is complete.

Chapter 3 – *Regulatory Issues* provides more detail on the U.S. regulatory requirements. For facilities not covered by PSM or RMP regulations, a broader application of the term pre-startup safety review is also addressed in this book.

1.5 THIS GUIDELINE'S AUDIENCE

This book is intended for anyone interested in developing a new PSSR program or upgrading an existing one. It can also help as a simple comparison tool to see where a mature program already uses some of the good practices or to help reinforce or improve upon the current methodology.

Whether a person is experienced or inexperienced in process safety, some of the typical positions in a company who may benefit from this guideline are:

- Managers of process safety management and PSM coordinators at a manufacturing facility
- Corporate process safety management staff
- Project managers and project team members whose projects initiate the need for a pre-startup safety review
- Engineers or other staff members performing management of change activities
- Operations, maintenance and other manufacturing personnel who may be part of a PSSR team
- Any employee participating in the PSSR program

1.6 HOW TO USE THIS GUIDELINE

Here are some suggestions on how the following chapters may be helpful to its audience.

CHAPTER 2 – What is A Pre-startup Safety Review?
- For workers with new PSSR team duties and for all facility managers, this chapter provides a description of the basic steps of pre-startup safety review.
- For site trainers and management, it gives guidance for training personnel involved in PSSR activities.
- For PSSR leaders and MOC owners, it discusses how scheduling aspects for different PSSR situations are addressed.

CHAPTER 3 – Regulatory Issues
- For safety professionals and management, this chapter provides an overview of the regulatory aspects of PSSR both in the U.S. and globally.
- For these readers, it also describes the specific U.S. OSHA and EPA regulations regarding PSSR.

CHAPTER 4 – A Risk-based Approach to Pre-startup Safety Review
- For all readers, this chapter introduces a key aspect for performing effective pre-startup safety reviews – using risk to determine the type or complexity level of PSSR to perform.
- For PSSR teams and PSM managers, it provides examples of tools some companies use to make these decisions.

CHAPTER 5 – The Pre-startup Safety Review Work Flow Process
- For all readers but especially PSM managers and PSSR leaders and team members, this chapter describes the components of a PSSR management system for a facility.
- For all readers, it describes considerations for implementing and following up on the steps in your customized PSSR program.

CHAPTER 6 – Methodologies for Developing Customized PSSR Checklist Items
- For PSM managers, PSSR leaders, team members, and employees participating in program development or upgrade, this chapter addresses the checklist, a common and very important tool for guiding effective pre-startup safety reviews.

- For personnel with PSM-related information technology duties or database management duties, it discusses aspects of computer-based electronic management of changes and pre-startup safety review programs.
- For users of the computer-based electronic MOC/PSSR systems, it discusses some typical features, benefits, and possible pitfalls associated with these MOC/PSSR system tools.

CHAPTER 7 – Continuous Improvement
- For PSM managers, PSM compliance audit team members and PSSR program developers, this chapter presents some approaches to maintaining a high quality PSSR program over time through self-assessment.
- For PSM managers, PSM audit team members and PSSR program developers, it offers some examples of typical compliance audit questions for internal and external audit teams to consider.

APPENDICES
- The appendices offer PSSR examples, a compilation of checklist questions, industry references, and regulatory references

1.7 REFERENCES

1-1 Occupational Safety and Health Administration, *Process Safety Management of Highly Hazardous Chemicals*, 29 CFR Part 1910, Section 119, Washington, DC, 1992.

1-2 Environmental Protection Agency, *Accidental Release Prevention Requirements: Risk Management Programs*, Clean Air Act, 40 CFR 68 Section 112 (r)(7), Washington, DC, 1996.

1-3 American Institute of Chemical Engineers, *Guidelines for Engineering Design for Process Safety*, Center for Chemical Process Safety, New York, NY, 1993.

1-4 American Institute of Chemical Engineers, *Guidelines for Implementing Process Safety Management Systems*, Center for Chemical Process Safety, New York, NY, 1994.

1-5 American Institute of Chemical Engineers, *Guidelines for Process Safety in Outsourced Manufacturing Operations*, Center for Chemical Process Safety, New York, NY, 2000.

1-6 American Institute of Chemical Engineers, *The Business Case for Process Safety Management,* edited by AntiEntropics, Inc. for the Center for Chemical Process Safety, New York, NY, 2003.

1-7 American Chemistry Council, *Resource Guide for the Process Safety Code of Management Practices*, Washington, DC, 1990.

2
WHAT IS A PRE-STARTUP SAFETY REVIEW?

In its most generic application to manufacturing, the term pre-startup safety review (PSSR) means the following:

> *PSSR is a formal review of a manufacturing process to verify that critical areas of the affected process have been assessed and addressed prior to using the process. Using the process could include: commissioning, introducing hazardous chemicals, or introducing energy.*

This chapter discusses the fundamental characteristics of PSSR, considerations for training personnel on PSSR as related to their roles, and special scheduling considerations.

2.1 THE BASICS OF PRE-STARTUP SAFETY REVIEW

There are eight basic steps to be considered for every type of pre-startup safety review program. Each facility can choose the best way to achieve their method of accomplishing the end goals of these steps. Those end goals may be:
- higher levels of process safety performance,
- better environmental risk management performance, and
- total manufacturing quality enhancements.

Consider how the steps offered below and the description of roles and responsibilities compare against your current practices.

2.1.1 Some Common Steps for Performing PSSR

Basic PSSR work-flow steps follow. An essential role for a facility's management is to ensure that knowledgeable people are available and assigned to provide the expertise for implementing the PSSR program. Thus, we begin with training.

Step 1 - Train the entire workforce on PSSR as related to their PSSR duties.

Typical job-performance related course titles might be:

- Basic awareness (typically provided as a part of a PSM awareness training module used during orientation of new employees)
- Management aspects (for those supervising any facet of the PSSR function, whether simply managing potential PSSR team members or managing program maintenance and upgrade)
- Potential PSSR team member training
- PSSR team leader training

Step 2 - Identify trigger events and determine if PSSR is to be performed.

A trigger event is any change being made to an existing process, any new facility being added to a process or facility, or any other activity a facility designates as such. One good practice is to define the specific trigger events (or their categories) for your facility in your program description. See section 2.4 for examples of non-change type trigger events. Typically, a pre-startup safety review will be performed before the startup of a new or significantly modified facility is authorized.

Here is a simple test to help determine when a PSSR is required for a triggering event at a facility:

- For PSM/RMP covered facilities:
    - A pre-startup safety review is required if the modifications to a facility are significant enough to require a change in the process safety information (PSI).
    - A pre-startup safety review is required for any non-PSM covered trigger event designated by the facility as benefiting from the review.
- For non-PSM/RMP covered facilities:
    - A pre-startup safety review is required for any non-PSM covered trigger event designated by the facility as benefiting from the review.

Very simply, a PSSR best practice is:
- for use when required by PSM/RMP
- for *replacement in kind* activities identified by the facility as a trigger event (that is, a maintenance job, design change, or repair item using original engineering design-based specifications and materials of construction or chemicals),

2. WHAT IS A PRE-STARTUP SAFETY REVIEW?   13

- for PSM- and RMP-covered facilities, PSSR can be considered a discrete project task required by regulation, and
- for startups after an emergency shutdown.

Step 3 - Determine the type of PSSR to perform – Simple/Short Form or Complex/Long Form.

This is where a facility may apply a qualitative risk-based approach to the PSSR. The first decision to be made is based upon the unique characteristics of the specific situation – the change or the new facility/major modification or other trigger event.

This process is described in detail in Chapter 4 – *A Risk-based Approach to Pre-startup Safety Review,* but in brief it is:

a) Determining whether the trigger event (the change) is straightforward, well understood, and can be signed off with use of minimal PSSR resources yet still be effective,

Or

b) Identifying a trigger event that is more complex and would benefit from a deeper engineering and process safety related check to provide an additional level of protection – a technique proven to improve safety performance.

Step 4 - Build the PSSR team.

Here is where your program defines its approach in regard to staffing requirements. Some facilities allow one to two-person teams for simple – short form – trigger events or changes and require larger teams for complex – long form – trigger events or changes. Chapter 5 – *The Pre-startup Safety Review Work Flow Process* provides more detailed suggestions.

Step 5 - Conduct the PSSR.

Again, the facility's PSSR procedure provides the details, but the following items are always applicable in some way:

- Schedule Team meetings. (One meeting may be sufficient for some PSSRs; multiple meetings may be needed for others.)
- Assign roles for the team members.
- Assemble PSSR compliance documentation or conduct field checks of physical activities required for the triggering event.
- Hold progress meetings as needed for the specific PSSR. (This is dependent upon the complexity of the triggering event.)
- Identify whether there are any pre-startup or post-startup action items and track them appropriately.
- When all pre-startup action items are resolved, obtain authorization to startup the change, new facility, modification, or process.

Step 6 - Complete the PSSR Documentation.

This is formally completing the hardcopy form or electronic documentation showing that the PSSR is complete, indicated action items are addressed and the post-startup items will be tracked.

Step 7 - Track any post-startup PSSR action items.

If there were action items to be completed after startup, follow their completion and documentation.

Step 8 - Seek continuous improvement in your PSSR program.

Use lessons learned from both the PSSR successes and setbacks your facility or industry experiences as events worthy of evaluation. These can help determine which programmatic aspects work – and which aspects don't work – in your PSSR management system.

2. WHAT IS A PRE-STARTUP SAFETY REVIEW?

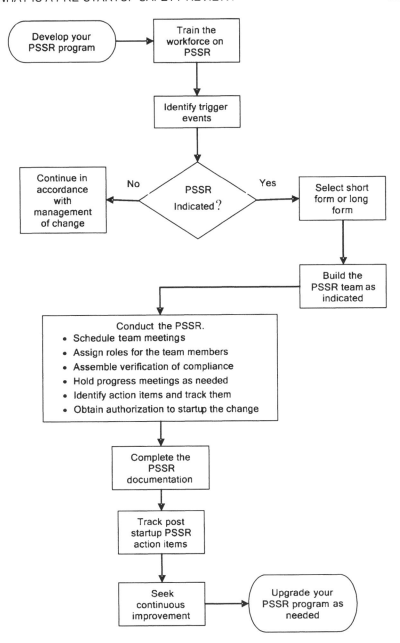

FIGURE 2-1 Basic PSSR work flow chart

2.2 WHAT IS A RISK-BASED APPROACH TO PSSR?

This book introduces the basics of applying a risk-based approach to pre-startup safety review. Specific aspects of trigger events may be examined for selecting the type of PSSR suitable for a given event.

This sub-categorization may involve evaluating the following aspects of each change or unit modification:

- Are there changes to equipment or trigger events in a process that involves especially hazardous materials regarding health, reactivity, and flammability or explosivity (for example, NFPA level 3 and above)?
- Are there new control systems or modifications that affect safety controls or interlocks (both safety & non-safety)?
- What is the project cost? For example, a facility may require the use of detailed PSSR for all affected projects over $10,000 – or any appropriate dollar amount for a specific facility.
- Does the change or trigger event involve a new type of equipment? For example, the first of its kind at the site or substantially different from older, similarly purposed equipment at the site.
- Are there multiple tie-in points to other systems or units? For example, one company chooses to use "three or more tie-in points" for a change or modification that affects the process safety information to indicate that a complex/long form PSSR approach will be used.
- If the process involved fails during or after startup, is it likely to result in a reportable health, safety, or environmental incident?
- Does the change or trigger event involve new or modified fire protection systems?
- Does the change or trigger event involve other new or modified life safety systems?
- Does the change or trigger event result from corrective actions taken due to a significant incident?

If any items are answered *yes*, in the above example of a risk-based approach, then study the change in detail and plan to perform a more complex PSSR rather than a simple PSSR. Additional considerations are provided in Chapter 4 – *A Risk-based Approach to Pre-startup Safety Review*.

2.3 THE ROLE OF TRAINING IN PRE-STARTUP SAFETY REVIEW

Your facility's business management, technical management, supervisory and staff personnel as well as each operations, maintenance, and support team member

2. WHAT IS A PRE-STARTUP SAFETY REVIEW?

should be familiar with their potential participation in implementing the practice of pre-startup safety review.

As indicated previously, some introduction to the concept and method of PSSR at a given facility is often presented as part of a process safety management system new employee overview. But two major categories of personnel may benefit from special consideration – the first category is your facility's PSSR leaders and team members and the second is your management team and the remainder of the workforce.

2.3.1 Training Team Leaders and Members

These trainees' roles indicate special attention to training design and implementation. Assuming a familiarity of PSM and PSSR from an orientation course or through day-to-day involvement in PSM activities, both the leaders and the members of PSSR teams need to know how to navigate though your specific PSSR program management system.

Both the leader and the members could benefit from training that includes the following learning objectives:

- Define PSSR as it applies to this facility.
- Discuss the benefits of performing pre-startup safety reviews from Chapter 1.
- Access the facility's PSM element administrative procedure from the facility document control system.
- Using the PSSR program PSM administrative procedure, walk through the steps for a mock change in a tabletop exercise.
- Given a change or trigger event, identify potential members for an associated PSSR team and explain why they were selected and what their roles might be.
- Given a change or trigger event, complete the PSSR documentation (electronic or hardcopy) using the results of the tabletop exercise.

Who is a likely candidate for acting as a PSSR team leader? Often, it is the person who initiated or is managing the MOC process for a change. Sometimes it may be the PHA leader for the change (if the change's risk exposure indicates a need for PHA). In reality, it could be any staff person, engineer, operator, maintenance team member, supervisor, or manager. The nature of the change or trigger event and its attending PSSR level of need is what may determine who the leader is for a specific PSSR. For large projects, consider choosing a leader who was not involved in the design or installation.

Pre-startup safety review team leaders would benefit from additional training on the following topics, some of which are available as public courses or possibly offered through your corporate organizational development group:

- *Process Safety Management* – Overview, current issues, and industry trends.

- *How to Facilitate Meetings* – Basic skills for preparing, leading and capturing the results from meetings.
- *Project Management* – A basic course using facility project management tools to schedule and complete activities to meet a goal.

2.3.2 Training Managers and the Remaining Workforce

Anyone involved in operating or maintaining a hazardous chemical process will most likely have been trained in an overview of process safety management, including pre-startup safety review. Some companies provide this overview to all employees at a chemical site to help ensure safety awareness at the facility. This level of training may be enough for the general workforce. However facility managers might benefit from additional training.

Managers of each operating department can benefit from attending the PSSR team training described above as the managers may actually act as team members or leaders and may have approval authority for the PSSR. A good exercise for any manager is to have them participate as a team member for an actual PSSR.

Another short interactive training session that may benefit managers would be a guided discussion among all the various managers at the facility of how their departments support PSSR and a brainstorm session for actions that will enhance that support. Some managers may need to discuss the scheduling aspects of personnel for major modification PSSRs. The facilitator for this guided session should capture the discussion topics and actions on a flip chart or LCD projector. These discussion-based training sessions can fill the role of a safety segment for a routine facility leadership team meeting. They also can sometimes result in unexpected manufacturing and safety enhancement suggestions.

2.4 SCHEDULING CONSIDERATIONS

Each change must be treated as unique for the management of change element to provide its greatest return on investment for a facility's process safety performance and its overall manufacturing reliability. Scheduling a PSSR can be a minor consideration for simple changes where the short PSSR form can be used. In some cases it may be a minor addition to the management of change work process. Other events may also trigger a PSSR according to facility policy. Scheduling PSSR for these events benefits from special attention.

Descriptions of these situations and some things to consider for each are provided in the following segments.

2.4.1 Capital Projects

When a major capital project involving a change or addition to a process safety covered facility is approved, the pre-startup safety review considerations typically begin. The length of time for design, site preparation, installation, and

2. WHAT IS A PRE-STARTUP SAFETY REVIEW? 19

commissioning may be months long. For major additions to a facility, the PSSR can essentially be treated as a sub-project of its own.

Consider the following items for capital projects:
- Can one complex PSSR cover this project adequately or should we plan multiple, more focused PSSRs to achieve the goal?
- Will the PSSR activities be carried out by the project team or by the operations and maintenance personnel who will own the equipment after the turnover?
- Will we have one team for the duration or separate teams for different portions of the PSSR activities?
- Will we perform the PSSR before commissioning or after?

2.4.2 Changes to Operating Facilities

Pre-startup safety reviews for changes to operating processes deserve special attention as well. These can be more routine changes, where PSSR can be done fairly quickly using the short form, or they can be changes as complex as that of a capital project and requiring the same level of scheduling and work assignments. The difference here is that there may be changes to or adjacent to operating chemical processes during the installation and pre-startup safety review activities. Look at the questions above concerning capital projects as well as the questions below when scheduling PSSR for these types of changes.

Consider the following items for operating facilities:
- Will the PSSR activities involve entering an operating area or will they be adjacent to an operating area? If so, plan for any special hazards the PSSR team must consider when performing their duties. Follow the facility's specific safe work practices.
- Are there multiple tie-ins to utilities or other processes? If so, plan to check that proper work permitting, especially hot work, hot tap, and line breaking safe work procedures have been followed.
- Will the PSSR activities need to address any special environmental release or waste handling issues due to this change? If so, plan to check each regulated environmental aspect of the change. This may include items such as release permits and waste manifests.
- Is there activity needed to write or upgrade procedures and other documentation?
- Was the training for the initial crew implementing startup of the change adequate and documented?
- Is there a plan to ensure all required training is performed in a timely fashion for all workers involved in operating or maintaining the changed operating facility before they operate or maintain it?

2.4.3 Temporary Changes

Temporary changes may also range from simple to complex in terms of pre-startup safety review considerations. Some multi product companies may even see each campaign for a new product to be a temporary change. They should consider all the guidance given in the previous paragraphs, but they have a unique characteristic.

For temporary changes, some companies have developed a practice of performing a PSSR for installing the temporary change, and then again at the change's removal. This practice helps ensure a smooth return to normal service and adequate communication or training if needed. Determine whether this practice is useful at your facility.

2.4.4 Restarting a Mothballed Process

As we are trying to describe pre-startup safety review techniques as they apply across all types of processing facilities, this situation shows how scheduling needs may differ across different types of plants.

The term "mothballed" in a general sense means a section of a unit or process has been out of use for an extended period of time. Some organizations use a specified period of time ranging from 2 weeks to 6 months as determined by the nature of their process. Here is where the scheduling consideration varies by the processes involved. The methods for properly mothballing (also called "laying-up") process equipment can be very different.

- Some facilities may simply isolate a process with valves, blinds, and flanges, then draw a vacuum or apply an inerting blanket. The restart may be straightforward.
- Some facilities may require special procedures and lay-up treatment materials and fluids to ensure a smooth return to operation. These procedures may be different based upon the anticipated duration of the mothball period.
- Some facilities have special startup procedures for processes coming out of an extended lay-up period no matter how simple the preparations for restart from mothballed conditions were.
- And yes, some facilities' units may have been mothballed but, for one reason or another, without following or maintaining proper lay-up procedures. These present special cases.

View the return to service of a mothballed unit as a change to be managed, and undergo an associated pre-startup safety review. The analysis of the proper startup from lay-up steps by working with the mothballed unit's restart team will allow the PSSR team to schedule their activities accordingly and more efficiently.

2. WHAT IS A PRE-STARTUP SAFETY REVIEW?

2.4.5 Post-turnaround Startup

Almost every chemical processing facility uses some form of the turnaround approach to allow major maintenance, detailed inspection, and equipment refurbishing. Typically, a turnaround involves a complete plant or unit shutdown and offers the chance to safely and properly install changes that may have been approved months ago but can best be performed during a shutdown period. There may be numerous improvements and enhancements to track as changes throughout the entire turnaround period.

Some questions to consider for restart after a turnaround are given below:
- Can working with the turnaround planning team in advance help the PSSR team(s) gain information to help schedule PSSR activities?
- Should we have an overall coordinator for all MOC/PSSR activities related to the turnaround?

A recent incident that occurred during a post-turnaround startup at a refining facility was investigated by the U.S. Chemical Safety Board (CSB) and the company involved. Both reports' recommendations imply that thorough pre-startup safety review implementation could have mitigated or completely avoided the incident.

2.4.6 Routine Maintenance

Although not required by regulation for OSHA PSM and EPA RMP covered facilities, some companies are finding that PSSR for some routine maintenance activities can provide a return on investment. Certainly process safety and personnel safety benefit, but general quality and reliability issues are also avoided with a good PSSR program.

Not all routine maintenance activities benefit from PSSR as much as others. For example, a company may choose to perform PSSR for a catalyst change, but may not choose to perform PSSR for routine pump replacements or rodding out heat exchanger tubes. The determination is open to facility management's discretion and is usually based on three aspects of the specific routine maintenance task. Scheduling a PSSR may need to be included in the work order. These are:
- Criticality to safe and efficient operation,
- Frequency of performance (frequent tasks being better understood), and
- Difficulty, whether physically difficult, intellectually demanding, or both.

2.4.7 Startup After Emergency Shutdown

Also not required by regulation for OSHA PSM and EPA RMP covered facilities (unless there were changes made prior to the restart), some companies are finding that PSSR for this critical operating phase has benefits to process safety performance. It is a self-defined trigger event in these cases.

Although an operating procedure should exist for this type of startup at all covered facilities, some processes are again found to be more complex than others.

There are certain processes that, due to their nature or automation, can be immediately restarted with minimal concern for upset. There are other chemical processes for which an emergency shutdown requires hours or days of preparations before they can be restarted. Here are some things to consider:

- Will a PSSR slow your restart process down? And if so, is slowing down a good thing or a bad thing? The time used to check PSSR-related activities may delay a simply restarted process, but slowing down for verification checks could actually help ensure a safe restart for more complex processes.
- Rather than do a formal PSSR, would it be wise to ensure final check steps for redundant verification of critical startup activities are built into the last section of the *startup after emergency shutdown* operating procedure?

2.5 REFERENCES

2-1 Occupational Safety and Health Administration, *Process Safety Management of Highly Hazardous Chemicals*, 29 CFR Part 1910, Section 119, Washington, DC, 1992.

2-2 Environmental Protection Agency, *Accidental Release Prevention Requirements: Risk Management Programs*, Clean Air Act, Section 112 (r)(7), Washington, DC, 1996.

2-3 American Institute of Chemical Engineers, *Guidelines for Implementing Process Safety Management Systems*, Center for Chemical Process Safety, New York, NY, 1994.

2-4 American Institute of Chemical Engineers, *The Business Case for Process Safety Management,* edited by AntiEntropics, Inc. for the Center for Chemical Process Safety, New York, NY, 2003.

2-5 U.S. Chemical Safety and Hazard Investigations Board, CSB Investigation Information Page, News Release dated October 27 2005, URL: www.csb.gov

3
REGULATORY ISSUES

Compliance with regulations is not only a legal necessity and an ethical duty for any corporation seeking to display social responsibility to its stakeholders, but compliance can also help ensure successful manufacturing business performance in the long term.

Process safety and environmental risk issues have matured as an international effort to share experience. Europe, Canada, South America, Australia, Asia, and the United States have all issued or proposed regulations that require programs for process safety related purposes. Industry groups supporting offshore oil exploration have also recognized the value of procedures to improve process safety. The International Organization for Standardization (ISO) includes the control and use of procedures as a key element of a company's quality management system and environmental management system. These influences affect more companies each year as they seek a global market for their products.

3.1 AN OVERVIEW OF PSSR INDUSTRY GUIDELINES AND REGULATIONS

In the United States, the impact of Occupational Safety and Health Administration (OSHA) regulation 29 CFR 1910.119 "Process Safety Management of Highly Hazardous Chemicals" and Environmental Protection Agency (EPA) 40 CFR Part 68 "Risk Management Programs for Chemical Accidental Release Prevention" is already being seen in industry performance. Developing, training on, and maintaining the required process safety work processes required by these regulations is a major effort.

Even before these regulations existed, industry groups, such as the American Institute of Chemical Engineers Center for Chemical Process Safety (AIChE/CCPS), the Institution of Chemical Engineers (IChemE), the American Chemistry Council (ACC), and the American Petroleum Institute (API), all promoted pre-startup safety review procedures as a sound management practice which can lead to improved process safety and process reliability.

Worldwide recognition of the benefits of process safety to workers and the environment became obvious beginning in the 1980s. Some foreign governments and industry groups adopted U.S. style process safety related standards or guidelines, continuing the sharing of information. Table 3-1 below lists organizations and an example document they produced to recommend or require pre-startup reviews, whether for safety, environmental, total quality purposes, or all three.

For the purposes of this guideline, we will primarily address the needs described in the U.S. OSHA and U.S. EPA descriptions of pre-startup safety review, as well as other good practices. A pre-startup safety review (PSSR) for U.S. OSHA PSM and EPA RMP compliance is required when any change modifies the process safety information.

This requires us to know several things:

- What a change is
- That there are two categories of changes – changes that do change the process safety information and changes that don't
- What the process safety information (PSI) is
- What a change to the PSI is.

So let us define a change. Very simply, it is an alteration to a process that is, not a direct replacement-in-kind. To be considered a replacement in kind, it must be an exact substitute for the originally designed equipment, chemical or technology.

3. REGULATORY ISSUES 25

TABLE 3-1
Global Examples of Pre-startup Safety Review Related Documents

ORGANIZATION	DOCUMENT
AIChE Center for Chemical Process Safety 1989	Guidelines for Technical Management of Chemical Process Safety
American Chemistry Council (ACC) September 1990	Responsible Care Code of Management Practices
Environmental Protection Agency Rule 40 CFR part 68 1996	Risk Management Programs for Chemical Accidental Release Prevention
Health and Safety Executive (HSE) United Kingdom 1984	A guide to the control of Industrial Major Accident Regulations, HS (R)21 (rev.)
ILO, International Labour Office (ILO) Geneva, Switzerland 1990	Prevention of major industrial accidents
International Organization for Standardization (ISO) 1995	Petroleum and natural gas industries - Health, Safety and Environmental Management Systems
Occupational Safety and Health Administration Rule 29 CFR 1910.119 May 26, 1992	Process Safety Management of Highly Hazardous Chemicals; Explosives and Blasting Agents; Final Rule
Organization for Economic Cooperation and Development 1992	Guiding principles for chemical accident prevention and response
Official Journal of the European Communities September 1985	Council Directive of 24 June 1982 on the Major-Accident Hazards of Certain Industrial Activities (the Seveso Directive)
United Nations Environment Programme (UNEP) 1996	Safety, health, and environmental management systems
World Bank 1985	Manual of Industrial Hazard Assessment Techniques
Korean OSHA PSM standard, Republic of Korea, 1996	Korea's Industrial Safety and Health Act - Article 20 Preparation of Safety and Health Management Regulations

Now we need to categorize those changes that require PSSR. It hinges on whether the change affects a certain set of data about the process – the process safety information. What is a change to the process safety information? It is any change to any of the following items that comprise the dataset that is the facility's PSI.

- Information pertaining to the hazards of the substances used or produced by the process (an effective MSDS program can typically address some of these items),
    o Toxicity information;
    o Permissible exposure limits;
    o Physical data;
    o Reactivity data:
    o Corrosivity data;
    o Thermal and chemical stability data; and
    o Hazardous effects of inadvertent mixing of different materials that could foreseeably occur.

- Information pertaining to the technology of the process,
    o A block flow diagram or simplified process flow diagram;
    o Process chemistry;
    o Maximum intended inventory;
    o Safe upper and lower limits for such items as temperatures, pressures, flows or compositions; and,
    o An evaluation of the consequences of deviations.

- Information pertaining to the equipment in the process.
    o Materials of construction;
    o Piping and instrument diagrams (P&IDs);
    o Electrical classification;
    o Relief system design and design basis;
    o Ventilation system design;
    o Design codes and standards employed;
    o Material and energy balances for processes; and
    o Safety systems (for example, interlocks, detection, or suppression systems).

What is an example of a change that would not change the process safety information? If you notice, operating and maintenance procedures are not listed. As long as a procedure changing how a process is operated or maintained does not require the written PSI documentation to change, it does not require a PSSR. Another example would be a change to piping that would change an isometric drawing, but not a piping and instrumentation diagram.

3. REGULATORY ISSUES

The questions the team uses during a PSM/RMP compliant PSSR should consider whether:

- **Construction and equipment meet the design specifications.**
  Obtaining field verification or performing document review for the new or modified process can validate design specifications for construction and equipment. If a change is not physical (such as a set point for an interlock shutdown), the method for the change and its anticipated effects should be reviewed.

- **Safety, operating, maintenance and emergency procedures are in place and adequate.**
  Examine the process management of change system documentation package for entries that indicate any safety, operating, maintenance or emergency procedures that were developed or revised for a modification. The management of change documentation package (as defined in the facility's management of change system) and referenced documents should be reviewed. Existing site safe work practice procedures should be checked to ensure they exist and are adequate.

- **A PHA has been performed for new facilities (or MOC was followed).**
  The management of change documentation packages and referenced documents should indicate when a Process Hazard Analysis was performed for the modification or new facility. The PSSR Team should verify all of the PHA recommendations have been implemented or otherwise resolved before the process can be judged ready to operate.

- **Training of each employee involved in the operating process is complete.**
  Typically, the management of change documentation package indicates when training of each employee involved in the startup of a new or modified process is complete. Training of other employees not directly involved with startup should be planned to occur before the first shift in which they would be required to operate or maintain the new or changed process. The PSSR Team may consider training complete for startup when this training information is verified.

- **Influential management personnel are involved in PSSR.**
  Involvement of influential management personnel in PSSR helps ensure that process safety is kept at an awareness level with the leadership at a facility.

3.2 BEST PRACTICES FOR PSSR

When an organization sees the added benefits to its process reliability and efficiency offered by the PSSR step, they sometimes choose to apply it in ways that go beyond external regulatory requirements.

Examples of a few of these good practices are provided below.

- **Regularly evaluating industry process safety related incident reports and how PSSR was potentially involved in the situation.** The PSM manager or coordinator at the site should research and share the experiences of other chemical processing facilities whenever it might apply to their facility.
- **Using electronic databases for capturing past PSSR documentation.** This allows PSSR teams to use search engines or other file indexing tools to evaluate similar past PSSRs. This encourages and enables taking advantage of the company's collective knowledge and lessons learned.
- **Performing PSSR on selected critical maintenance activities** even when the PSSR step may not be required by regulatory standards. This practice provides a double check on key performance issues.
- **Involving many different workers in the PSSR process.** This promotes reinforcement of the process safety program at the facility and provides documented employee participation.
- **Selecting a PSSR team leader who is somewhat removed from the specific project involving the change.** This helps remove the possibility that project schedule pressures and pride of ownership will negatively influence the review.
- **Showing open management support for the importance of PSSR.** By seeing facility management personnel occasionally delay a planned startup to ensure the final details required by the PSSR team are fulfilled before authorization for startup, all employees realize the critical nature of the review.

Again, these are not driven by requirements; they are good practices. Evaluate your program to see if there are other good practices your facility uses which may be beneficial to your pre-startup safety review performance. Share these with other facilities as appropriate.

The next two sections consider other regulatory areas where PSSR may assist with compliance.

3.3 ENVIRONMENTAL CONSIDERATIONS

There are two main points to be made about environmental considerations. One is related to the task of installing the change or implementing the trigger event under review. The other is related to operating and maintaining the system with the new change or trigger event in place.

For activities surrounding the installation of the change, the facility work permitting program should have made provisions for special environmental concerns:

- Special waste generation aspects
- Special hazardous material storage during the installation
- Considerations for chemicals that may be involved in the preparation or installation/cleanup phase

The pre-startup safety review process can check that these items were evaluated and in place.

With regard to long term operation of the changed process, consider the following environmental characteristics:

- If new chemicals are being introduced, are there specific environmental issues to address?
- If waste characterization for the process is altered, have revisions to permits and manifests been considered?

The PSSR process can verify the regulatory needs are met prior to exposing the facility to the risk of findings or citations.

3.4 GENERAL SAFETY, SECURITY, AND OCCUPATIONAL HEALTH CONSIDERATIONS

Although the definition of process safety primarily involves preventing catastrophic incidents, basic worker safety and health receives attention as well when building a rigorous integrated process safety management program.

As with environmental considerations, the two periods of concern are during installation and later operation and maintenance of the change.

For activities surrounding the installation of the change or other trigger event, again the primary level of protection is the facility work permit system. Special personnel protective equipment (PPE) will be called out for the hazards and risks associated with the installation.

For activities related to long-term operation and maintenance, the PSSR can check that any new PPE or exposure concerns are addressed.

Security issues may not be crucial to the chemical engineering aspects of a change or trigger event, but they may take on the role of a necessary administrative or management consideration before the change can be implemented. For example, some raw or intermediate materials may be governed by agencies similar to the U.S. Bureau of Alcohol, Tobacco and Firearms (ATF), or the Chemical Weapons Convention (CWC). Some processed materials may have black market value and be subject to pilfering, and some materials may be subject to internal company control due to their hazard exposure or risk ranking. Unfortunately, some facilities may be considered potentially attractive targets for terrorist activity. Consider whether issues such as lighting, monitoring, security barriers, personnel ingress and egress systems, or other loss prevention consideration must be evaluated prior to startup. This could indicate a revision to your security vulnerability analysis (SVA).

3.5 REFERENCES

3-1 Occupational Safety and Health Administration, *Process Safety Management of Highly Hazardous Chemicals*, 29 CFR Part 1910, Section 119, Washington, DC, 1992.

3-2 Environmental Protection Agency, *Accidental Release Prevention Requirements: Risk Management Programs*, Clean Air Act, Section 112 (r)(7), Washington, DC, 1996.

3-3 American Institute of Chemical Engineers, *Guidelines for Implementing Process Safety Management Systems*, Center for Chemical Process Safety, New York, NY, 1994.

3-4 American Institute of Chemical Engineers, *Guidelines for Managing the Security Vulnerabilities of Fixed Chemical Sites*, Center for Chemical Process Safety, New York, NY, 2003.

4
A RISK-BASED APPROACH TO PRE-STARTUP SAFETY REVIEW

Many chemical processing facilities use risk modeling to prioritize or rank the urgency or resource levels required for initiating certain actions. Risk ranking is also useful for examining the results of not taking an action or delaying it. Effective pre-startup safety reviews (PSSR) are supported by using a similar risk-based approach to make some basic decisions for planning and implementing an appropriate application of effort to the PSSR for any trigger event.

4.1 USING RISK ANALYSIS TECHNIQUES TO SELECT THE LEVEL OF DETAIL FOR A PSSR

What do we mean by a *risk-based* approach to pre-startup safety review? This term indicates that we will use the performance based aspect of the PSSR regulations, accepted engineering methods, and consensus-based industry guidance to more efficiently and effectively design each trigger event's PSSR activities. The goal is to make the best use of organizational resources based upon the risk attributed to the changed process or trigger event.

This guideline offers several examples of qualitative tools to help a PSSR leader and his or her team determine whether a PSSR can be done simply (for straightforward trigger events) or whether it must be handled in a more complex manner. Even before a change gets to the point where you ask, "Is it a simple or complex PSSR?", risk has been weighed for the change. Think about a basic management of change program:

- First someone saw a potential change or trigger event and weighed its risk-based nature upon whether or not it met the definition for replacement-in-kind or a change.
- Next, if it was a change, someone evaluated whether the change's risks as designed were enough to require that a process hazard analysis (PHA) was performed for the change

- Finally, if it was a change, someone evaluated the risk of the change by measuring a secondary source of information – the PSI. That secondary source is whether or not the change, if implemented in concert with process safety tenets, will cause changes in the process safety information. There is a risk that the process safety information integrity erodes with every poorly documented or improperly managed change, and as the changes become riskier themselves, it is more likely that the process safety information will be affected. It is in a sense, a tell-tale for the risk level.

There are several levels of risk analysis within most methodologies for assessing risk: quantitative, semi-quantitative, and qualitative. For PSSR concerns we deal almost exclusively with qualitative assessments, that is, just a determination of high or low risk. Generally any truly quantitative risk analysis (QRA) indicated for a trigger event would be performed to enhance the process hazard analysis. The associated PSSR for such a trigger event would simply follow action item progress related to the quantitative risk assessment's action items. In this case the PSSR helps assure that any action items from a QRA are appropriately followed.

Once we have decided that a PSSR is needed to meet our facility's requirements, there are actually three points in the decision making processes for PSSRs where risk-based applications can apply.

- The first is the simple versus complex PSSR decision point.
- The second is at the design phase of a more complex PSSR.
- The third possibility for applying risk is when scheduling action items that must be followed up after startup or those that must be completed before startup.

4.1.1 A Case of Complexity Versus Simplicity

At the first decision point we encounter in performing effective PSSRs, we evaluate the overall risk presented by:
- the act of installing a change or other trigger event and
- the effect of operating with the change in place.

This risk-based ranking is fairly straightforward. Risk will be assessed for the activities associated with the trigger event or installing the change, introducing chemicals to the changed equipment, the startup period, as well as the equipment's operation after startup.

It is impossible, and essentially unwarranted in many cases, for facility management to resolve all process change issues simultaneously and at the highest level of detailed attention. Practicality and pragmatism tell us that some items are

4. A RISK BASED APPROACH TO PRE-STARTUP SAFETY REVIEW

less urgent or risk-laden than others. Therefore, we must allocate our resources for both level of effort and scheduling in the most effective manner.

The following example of a basic priority risk-ranking matrix system is provided to display one tool for determining whether the change or trigger event is simple – and able to use a short form PSSR – or more complex, thus in need of a long form PSSR.

The term *simple* should not be misunderstood as we use it in this book. It does not mean this category of PSSR is less important than a complex PSSR. It is intended to indicate the level of effort needed is well understood and fairly straightforward. Whenever we use the terms *simple PSSR* or *short-form PSSR*, they indicate a less resource-intensive approach to verifying readiness for startup when the trigger event has a lower level of risk.

The term *complex* merely indicates that the PSSR will require some level of special planning and effort due to its unique risk-based characteristics or novelty to the site. Whenever we use the terms *complex PSSR* or *long-form PSSR*, they indicate a more customized and sometimes more time-consuming approach to verifying readiness for startup. We will see this as we examine PSSR team makeup and review planning examples.

One tool used for risk ranking is the graphically organized matrix chart as shown below:

Instructions: • Track the likelihood against the severity • If risk ranking is 3 – use the short PSSR form • If the risk ranking is 1 or 2, use the long PSSR form	3 LOW Event which could cause a small release of a highly hazardous chemical and/or result in an easily controllable fire. Examples of low severity events would be those which could result in only minor employee or contractor injuries and minor property damage (< $500M).	2 SERIOUS Event causing significant release of highly hazardous chemicals, a fire, or an explosion which presents the potential for moderate danger to employees, the public, or workplace. Examples are one or more employee or contractor injuries, or large equipment damage ($500M to $5MM).	1 CATASTROPHIC Event causing a major uncontrolled emission, fire, or explosion involving one or more highly hazardous chemicals, that presents serious danger to employees, the public, or the workplace. Examples of catastrophic danger might include one or more employee or contractor fatalities, public injuries, or extensive property damage. (> $5MM)
1 HIGH Exposure to this event can be very frequent. For example, tasks which are performed on a daily or weekly basis, conditions that continuously exist, or equipment deficiencies.	Risk Ranking 2 Medium	Risk Ranking 1 Higher	Risk Ranking 1 Higher
2 MODERATE Exposure to this event can be infrequent. For example, tasks that are performed no more frequently than monthly or less during the course of the operating year	Risk Ranking 3 Lower	Risk Ranking 2 Medium	Risk Ranking 1 Higher
3 LOW Exposure to this event can be rare. For example, tasks which are performed less frequently than once/year up to every three or four years for turnarounds or things that almost never happen	Risk Ranking 3 Lower	Risk Ranking 3 Lower	Risk Ranking 2 Medium

INCREASING LIKELIHOOD OF THE EVENT ↑

INCREASING SEVERITY OF THE EVENT →

FIGURE 4-1 An example risk matrix chart

Some companies use different levels of frequency and severity and others may also look at other dimensions separately like capital expenditure levels. Often, the simple PSSR is reserved only for lower risk ranking. Medium to high would indicate the need for a complex PSSR.

Another tool used to make this simple PSSR versus complex PSSR risk-ranking decision is a risk questionnaire. In this case, a facility applies a set of standards against which each PSSR is evaluated. It may be performed:

- as a series of steps in the PSSR administrative procedure as in the example offered in Table 4-1 – *Typical Risk-based Decision Making Steps in a PSM PSSR Procedure*,
- by keying the information into a set of data fields in an electronic database system, or
- by documenting on a separate hardcopy or electronic form.

TABLE 4-1
Typical Risk-based Decision Making Steps in a PSM PSSR Procedure

If the change management review indicates the process safety information must be changed, ask the following questions about the change to determine if it is classified as a simple or complex PSSR:

- For Simple PSSRs use form FRM1 – *PSSR Short Form*
- For Complex PSSRs use form FRM2 – *PSSR Long Form*

When any one of the following are true statements, use the *PSSR Long Form* to plan and implement the review:

- The change involves equipment in a process that involves materials with:
  - health ratings of 3 or 4,
  - reactivity ratings of 3 or 4, or
  - flammability ratings of 3 with operating temperatures above 75 degrees F or operating pressures above 15.2 psig and
  - all materials with flammability rating of 4.

- The project cost is greater than $10,000.

- The change involves a new type of equipment (that is, the first of its kind at the site or substantially different from other equipment on site).

- The change involves three (3) or more tie-in points.

- If the process involved in the change fails, it is likely to result in a reportable incident for the facility.

- The change involves new control systems or modifications that affect safety controls or interlocks

- The change involves new or modified fire protection systems.

If none of the statements above are true, the team may elect to use form *PSSR Short Form* to plan and perform the PSSR.

4. A RISK BASED APPROACH TO PRE-STARTUP SAFETY REVIEW

Some companies choose to use the complex PSSR form itself as the method of risk ranking. In this case, they have each PSSR team evaluate every specific item on the checklist for applicability for every trigger event, even when it is an obviously simple modification. In this way, the team qualitatively addresses risk for each item they include in the review.

4.1.2 The Term Complexity Includes Novelty

A third point at which risk-based analysis comes into play for pre-startup safety reviews is in designing the review for a novel process or trigger event. As a renowned process safety expert and author Trevor Kletz has pointed out, many catastrophic accidents have occurred because:

Personnel did not know what they did not know.

The starting point in developing the review plan for a novel trigger event is similar to a complex change or trigger event; however, the degree of novelty should also be assessed. The degree of novelty is important because it is a measure of how likely past experience can be used to predict future performance. The following guidelines offer one example of how to classify the degree of novelty.

High Novelty:

- Does the trigger event involve new chemicals or ones previously unused at the site?
- Does the trigger event involve new processes or ones previously unused at the site?
- Does the trigger event involve a new type of equipment or equipment previously unused at the site?

Medium Novelty:

- Does the trigger event involve existing chemicals used in a new or unusual way?
- Does the trigger event involve an existing process operated outside of the normal range of process variables (for example, operated differently for a plant trial)?
- Does the trigger event involve existing equipment used for a new or unusual service?

Low Novelty:

- Are existing chemicals, process, or equipment used within normal or expected operating conditions for this trigger event?

4.1.3 The Effect of Complexity on PSSR Team Size and Expertise

Once we have determined whether the change or trigger event needs a simple or complex pre-startup safety review, we need to address team size and the types of subject matter expertise we need. Chapter 5 – T*he Pre-startup Safety Review Work Process* addresses general team selection steps.

In regard to how complexity affects the PSSR team composition, the selection of the PSSR short-form approach using a risk-based approach means a minimum team is required.

For example, a two-person team, the PSSR team leader and one member, may be all the signature authority needed to sign off on a PSSR short form if signatures are required by your site policy. An example form is shown in Table 4.2 – *Example short form for lower risk / simple PSSR*.

Typical two-person teams for PSSR short-form category changes are made up from a combination of:

- The management of change (MOC) initiator (which can be any position),
- An operations team member,
- An engineer or technical staff member,
- A safety team/process safety team member, or in some cases
- A management team member,
- A maintenance team member,
- A contracted position involved in the change.

The typical effort for a simple PSSR is for review of the documentation associated with the change or trigger event, a physical walkthrough of the change or other verification of proper installation, then hardcopy or electronic signatures allowing any other preparations in accordance with plant policy or regulatory requirements for startup to proceed.

An example change might be – *changing tube materials on a shell and tube heat exchanger.* In this example, the materials will have been researched for compatibility with original design specifications before the change was initiated and the change-out procedure itself will likely be straightforward and done in a safe-process mode using proper energy control procedures. Operating and maintenance procedures will most likely not be affected too severely by such a change, and standard safe work practices will be sufficient for the installation work and future long-term operation of the equipment.

4. A RISK BASED APPROACH TO PRE-STARTUP SAFETY REVIEW

A two-person team could easily check that PSSR required documentation of the change or trigger event is in evidence and visit the worksite or review maintenance department job information. If they are satisfied the work is complete and in accordance with the change package, they can authorize the pre-startup safety review as complete.

Team makeup for more complex changes or trigger events – those requiring the PSSR long-form approach – is potentially unique for every instance. Depending upon the trigger event's nature, some expertise to consider adding to the pre-startup safety review team includes:

- Chemists
- Civil or structural engineers
- Construction department personnel
- Contractor participants
- Environmental scientists or specialists
- Fire and emergency response team/fire chief input
- Human resources representatives
- Industrial hygienists
- Lab personnel
- Maintenance personnel/engineers
- Materials engineers/metallurgists
- Original equipment manufacturer (OEM) representatives – a factory or team services engineer
- Process control (electrical/instrumentation) engineers
- Purchasing or stores personnel
- Quality assurance specialists
- Recently retired employees with pertinent knowledge, skill, or experience
- Research technical personnel
- Safety/process safety professionals
- Technical consultants or equipment specialists

The typical effort for a complex PSSR is never typical. If a trigger event, change, or new facility is a large enough project or a major technological switch for the site, with high business importance or long installation duration, there may be numerous points to review documentation associated with the change, several physical walkthroughs of the change in progress (and maybe by different types of teams), then hardcopy or electronic signatures to show the entire team authorizes PSSR is complete and startup may commence. An example of a more complex change form is shown in Table 4.3 – *Example long form for higher risk / complex PSSR*

An example complex change might be – *adding a new chemical to the process using an automated injection system*. In this example, the materials will also have been researched for compatibility with original design specifications before the change was initiated – but as this is a new chemical, the PSSR team should review the materials for initial compliance, verify the MSDS system has been updated, and possibly check on the design and installation of the injection-system equipment during construction. Operating and maintenance procedures will most likely need revision or development and safe work practices may need to be revised to reflect new PPE requirements or chemical hazards associated with the trigger event.

There are more complex changes and trigger events than that offered in the example, such as a unit expansion or major modification, but the same considerations need to be made – What is the specific nature of the trigger event? What are the risks? And which team members will best help avoid or mitigate those risks? As a thought exercise, consider a non-change trigger event that merits a complex PSSR, such as startup after an emergency shutdown.

4.1.4 The Effect on the Level and Scope of the Review

The second point at which risk-based analysis comes into play for pre-startup safety reviews is in designing the review for a complex change or trigger event. The starting point in developing the review plan is in reviewing the facility's PSSR long-form checklist items. Look through your facility's checklist in regard to the trigger event under review and determine which PSSR items are applicable.

The higher the qualitative risk level associated with the trigger event, the more likely that at least one extremely critical PSSR item will present itself as an item worthy of review.

In addition, a larger project will result in a larger number of PSSR items being considered applicable. That means more PSSR team action items to resolve.

Also, a larger complex change or trigger event may occur over a period of weeks or months, and the PSSR may benefit from scheduling checks that match the installation progress.

4.2 A DECISION GUIDELINE FOR DESIGNING A PSSR

The following sections assist in establishing a risk-based decision guideline for pre-startup safety reviews.

4.2.1 A Definition of Risk-based PSSR – A Qualitative Approach

Based upon the descriptions in previous chapters, we know the definition of a pre-startup safety review is:

A final check, initiated by a trigger event, prior to the use or reuse of a new or changed aspect of a process. It is also the term for the OSHA PSM and EPA RMP

4. A RISK BASED APPROACH TO PRE-STARTUP SAFETY REVIEW

element that defines a management system for ensuring that new or modified processes are ready for startup.

But more descriptively, it is a method for helping ensure that processes are ready for startup by verifying aspects such as:

- Construction and equipment are in accordance with design specifications.
- Safe work practices, operating procedures, maintenance procedures, and emergency response procedures are in place and are adequate.
- A process hazard analysis has been performed (if needed) and recommendations have been resolved or implemented, and modified facilities meet MOC requirements.
- Training of each employee involved in operating a process has been completed.

A risk-based PSSR, therefore, is a PSSR that focuses review energies on the four dimensions of PSSR (design / construction specifications, all procedures, process hazard analysis action items, and employee training) *in relation to the risk of not verifying the specific issue.*

4.2.2 An Example Algorithm

An algorithm is essentially a procedure or a set of well-defined instructions for accomplishing a task. Applying an algorithm to a process effectively depends on accurate data and its analysis. That is where human factors come into play. Helping everyone in your organization understand their role in pre-startup safety reviews will result in better operational discipline in regard to meeting your PSSR compliance needs.

The basic algorithm steps below describe performing a risk-based PSSR. They will be expanded upon in Chapter 5 – *The Pre-startup Safety Review Work Flow Process.*

- Become aware of a triggering event (a change in regard to process safety or other organizationally defined standard).
- Assess the event's need for management of change and pre-startup safety review.
  o Not a replacement in kind = a change
  o Change affects process safety information = needs a PSSR
  o Change or event meets other internal non-PSM/RMP criteria for PSSR = needs a PSSR

- Apply a qualitative risk-based analysis to the event to determine the event potential severity and likelihood.
    - A risk matrix
    - A risk questionnaire
    - Another method of assessing risk
- Assess the PSSR event in regard to its complexity using the risk-based analysis results.
    - Simple = Implement the PSSR short form
    - Complex = Implement the PSSR long form
- If a simple PSSR is indicated:
    - assign team,
    - schedule document reviews and physical reviews,
    - follow up on action items,
    - document the equipment is safe for startup using the PSSR short form when all criteria are met.
- If a complex PSSR is indicated:
    - evaluate the change or trigger event against the PSSR long-form review items and determine which items apply,
    - expand upon PSSR items as needed,
    - design the team makeup based upon expertise indicated by the event,
    - assign team members,
    - schedule document reviews and physical reviews as needed throughout the installation phase,
    - follow up on action items,
    - document the equipment is safe for startup using the PSSR long form when all criteria are met.
- Follow any post-startup action items to completion.
- Verify the documentation is maintained in accordance with the facility's document-retention guidelines.
- Audit the overall PSSR program performance periodically.

Check this algorithm against your facility's PSM administrative procedure for pre-startup safety review, management of change, and compliance audits. Are the basic steps present in some way? Evaluate whether your PSSR program could benefit from update to reflect these items more clearly.

4. A RISK BASED APPROACH TO PRE-STARTUP SAFETY REVIEW 41

4.3 TYPICAL CONSIDERATIONS FOR ALL PRE-STARTUP SAFETY REVIEWS

Using a risk-based ranking system to plan workload and resource requirements for pre-startup safety reviews is an excellent tool, but even the PSSRs for trigger events with lesser risk values have things in common with more complex trigger events.

Notice that the short-form and long-form PSSR checklist examples both address the following topics, but with varying levels of required effort depending upon the risk-based assessment.

4.3.1 Hardware and Software: Equipment, Instrumentation, and Process Control

For trigger events that involve equipment or its control devices, there is typically a segment of a complex PSSR form that outlines the many categories of variations which may apply.

Some hardware changes or trigger events can be simply verified as the installation progresses. A PSSR review of a hardware item might be as simple as checking the work permit and walking into the unit to verify the piping, vessel, valve, or other item is in place and matches the specifications.

For complex changes or trigger events, the total modification involving equipment or instrumentation may include anywhere from a handful to a hundred or more different physical items being installed. When appropriate, the PSSR team should physically verify (or document others have physically verified) that they are installed as designed.

Software changes can be trickier to verify. In many cases, a software change affecting process safety information can be made without any explicit observable physical change until it is implemented. How do you check such a change or trigger event prior to use in these cases? Some facilities simply ensure the change is designed to result in a fail-safe mode. Others design a test for the loop to ensure that the signal is as intended. Software may be able to be tested on simulators in some cases.

4.3.2 Documentation: Process Safety Information, Procedures, and Maintenance Management System Data

In general, the more complex a change or trigger event is, the more likely it is to have higher risk rankings, and the more likely it is to affect the PSM-related documentation at your facility.

This means that the process safety information needs to be revised as documentation of the changed process, and there needs to be documentation that it was checked (for example, the PSSR form or database) to document that process safety was considered. Examples of piping, equipment, or automation changes that will most likely affect the PSI include:

- Any P&ID change, such as a new branch, a new valve, or different type of valve
- A new control loop or control logic (even if existing sensors, controllers, and final control elements are used)
- Changing pipe diameter
- Adding or removing a filter, instrument, or anything attached to a piping system in hazardous chemical service
- Changing materials of construction
- Changing impeller size on a pump
- Changing the seal design on a pump
- Changing the control logic for an interlock (safety & non-safety)
- Changing media in a fire suppression system
- Changing the ductwork, air mover, filtration unit, or any component of a ventilation system

4.3.3 Training: Quality and Verification of Completeness

For the purposes of PSM and RMP compliance, any new or revised versions of the following documents are considered impetus for training affected personnel:
- Process overviews
- Safe work practices
- Operating procedures
- Maintenance procedures
- Emergency response procedures

The more complex (or novel) the PSSR, based upon the size of the project or the risk-based assessment, the more likely it is that one or more of these documents needs to be revised.

For some procedure changes, a simple review of the change with the affected personnel can be enough. For other changes, a formal training session or walkthrough may be appropriate. Here is where the training staff at your site might get involved in the PSSR process. It can be argued that ineffective training is worse than no training at all. If a facility is PSM or RMP covered, there is a paragraph in each rule implying a requirement to verify that the training was understood. Whether regulated or not, it will pay off in the long run when training design quality is appropriate for the activity.

Identifying the required or desired training audience is also critical. One issue with many PSSR programs is that the initial shift of workers involved in the starting the process after a trigger event receives the training, but later shifts and persons on vacation or other leave may get overlooked.

4. A RISK BASED APPROACH TO PRE-STARTUP SAFETY REVIEW

Facilities should consider establishing their own criteria for judging whether training has enhanced a person's skills or knowledge as related to their job performance, which ultimately includes safety and process safety dimensions. Different methods can be used when training is desired or required for a trigger event. For example, you could use any of the following methods, but your selection should be based upon good instructional systems design rules to fit the specific nature of the item:

- Written exams
- Field demonstrations
- Tabletop reviews of the item at shift change

The risk level assigned to the trigger event may impact the type and intensity of the training delivered. The essential method is to have a verifiable, objective means of documenting understanding.

4.3.4 Special Items: Specific Safety, Health, and Environmental Issues

Every trigger event is important enough to pay special attention to a major reason the review exists. That reason is safety, health, and environmental performance.

The items that are looked at here include:

- General personnel safety
- Machinery safety
- Ergonomics
- Occupational health and industrial hygiene
- Environmental permitting or release prevention

For example, a trigger event for PSSR could easily introduce new chemical hazards, new permitting requirements or permit update triggers, training requirements, or prevention plan updates. Consider these aspects of the work during the PSSR.

4.4 AN EXAMPLE RISK-BASED QUESTIONNAIRE

Below is an example of a risk-based questionnaire system. Several more will be offered in later chapters and the appendices.

As you can see, these examples and others can be completed in hardcopy or electronic format. Many companies have electronic database systems for tracking the PSSR activities and the action items.

Note that the short form is essentially the first page of the long form, which is further customized based upon the activity. For PSM and RMP covered facilities similar words should appear on any PSSR documentation performed for a covered process in order to confirm the process is ready for startup.

TABLE 4-2 Example short form for lower risk / simple PSSR

INSTRUCTIONS: Complete the form with basic information for the review. If any special PSSR activities are indicated by the team, add additional pages as necessary. The PSSR team leader should track each applicable item as it is completed. When all items required prior to startup are complete, obtain the team member signatures. Transmit the PSSR completion to the party responsible for startup.

PRE-STARTUP SAFETY REVIEW - SHORT FORM			
Date:	PSSR Team Leader:		
Facility/Process Equipment Reviewed:			
Type of Startup (Check one)	New Construction:☐	Modified Process/Restart:☐	
Recommendations: (Expand as needed. Attach relevant documents to this form)			
Item No.	Description	Initials	Date Resolved
PSSR Completion Summary: The following issues have been resolved and the undersigned believe the process/facility is ready for startup.			
• The construction and equipment meet design specifications.			
• Safety, operating, maintenance, and emergency procedures are in place and are adequate.			
• For new facilities, the initial process hazard analysis (PHA) has been performed and recommendations have been resolved.			
• Training of each employee involved in the operating process is complete.			
• Changes made to modify the process/facility have been reviewed and authorized by the facility management of change program.			

Confirmation by PSSR Team Members		
Name/Title	Signature	Date

4. A RISK BASED APPROACH TO PRE-STARTUP SAFETY REVIEW 45

TABLE 4-3 Example long form for higher risk / complex PSSR

INSTRUCTIONS: Complete Part 1 with basic information and retain it for later signature by the team. Use Part 2 to identify and plan PSSR items of interest and activities. Mark N/A for PSSR items in Part 2 which are not applicable for this trigger event. Identify the person responsible and the projected completion date for all applicable items. If any special PSSR activities are indicated by the team and are not reflected on this form, add additional pages as necessary. The PSSR team leader should track each applicable item, and as it is completed, initial its closure. When all items required prior to startup are complete, obtain the team member signatures on Part 1. Transmit the PSSR completion to the party responsible for startup.

PART 1 - PRE-STARTUP SAFETY REVIEW - LONG FORM			
Date:	PSSR Team Leader:		
Facility/Process Equipment Reviewed:			
Type of Startup (Check one)	New Construction:☐	Modified Process/Restart:☐	
Recommendations: (Expand as needed. Attach relevant documents to this form)			
Item No.	Description	Initials	Date Resolved
PSSR Completion Summary: The following issues have been resolved and the undersigned believe the process/facility is ready for startup.			
• The construction and equipment meet design specifications.			
• Safety, operating, maintenance, and emergency procedures are in place and are adequate.			
• For new facilities, the initial process hazard analysis (PHA) has been performed and recommendations have been resolved.			
• Training of each employee involved in the operating process is complete.			
• Changes made to modify the process/facility have been reviewed and authorized by the facility management of change program.			
Confirmation by PSSR Team Members			
Name/Title	Signature	Date	

Item	Responsibility	Projected Completion Date	Completion Date	Initials
PROCESS HAZARD ANALYSIS ISSUES:				
Pre-startup action items closed				
Post-startup action items ID'd				
Other:				
DOCUMENTS IN PLACE:				
Operating Procedures				
Maintenance Procedures				
Safety Procedures				
Emergency Response Procedures				
Other:				
PROCESS TECHNOLOGY INFORMATION:				
P&IDs				
Process Flow Diagrams				
Safety Equipment Plot Plan				
Area Electrical Classification				
MSDS/Chemical hazards				
Safe Operating Limits				
Process Description/Chemistry				
Equipment Lists				
Ventilation Systems				
Heat & Material Balance				
Relief Valve Summary List				
Instrument Safety System List				
Mechanical Safety System List				
Piping Schedule				
Maximum Intended Inventory				
MECHANICAL INTEGRITY INFORMATION:				
Rotating Equipment Records				
Electrical Equipment Records				
Inspection Records				
Relief Valve Records				
Equipment Records				

4. A RISK BASED APPROACH TO PRE-STARTUP SAFETY REVIEW   47

Item	Responsibility	Projected Completion Date	Completion Date	Initials
Equipment Specifications				
Electrical 1 line drawings				
INSTRUMENTATION & CONTROLS INFORMATION:				
Advanced Controls Documentation				
DCS or PLC Documentation				
DCS Screen Revisions				
TRAINING: (for startup personnel at a minimum)				
Operator/Maintenance Process Overview Training				
Operator Procedure Training				
Emergency Shutdown Procedures				
Emergency Evacuation Procedures				
Maintenance Procedure Training				
HAZCOM Training Certified				
Fire Suppression System Training				
Fire Department Training				
Emergency Response/Hazmat Training:				
Level 1: Awareness				
Level 2: Operations				
Level 3: Technician				
EPA Hazardous Waste Training				
Safety Shower/Eye Wash Training				
GENERAL FACILITY REQUIREMENTS				
Construction/Equipment Meet Design Specifications				
Ladder/Platforms Meet OSHA Standards				
Handrails/Toe boards/Walkways Meet OSHA Standards				
Electrical Systems In Service				
Lighting Adequate				

GUIDELINES FOR PERFORMING EFFECTIVE PRE-STARTUP SAFETY REVIEWS

Item	Responsibility	Projected Completion Date	Completion Date	Initials
Safe Access to Block Valves, gauge glasses, and Bleeders				
Temporary Electrical Removed				
Electrical Meets OSHA Code				
Fire Protection Systems Tested				
Fire Alarm Systems Working/Tested				
Fire Extinguishers In Place				
Fire Hydrants & Monitors In Place				
Fire Hoses In Place				
Fire Water Pumps Operational				
Oxygen/Gas Monitors/Alarms Calibrated				
Utilities Marked in accordance with OSHA Standard				
Identify Piping System Dead legs and Eliminate if Possible				
Containers and Piping Marked in accordance with OSHA Standard				
Safety Shower/Eye Wash Operational				
Gas Rescue Equipment In Place				
Steam Systems In Place/Operational				
Control Loops Checked/Operational				
Steam Tracer Circuits Active				
Steam Lines Insulated in accordance with OSHA Standard				
Steam Vents & Pressure Relief Valves Operational				
Instrument Air Systems Operational				
Instrument Air Dryer Systems Operational				

4. A RISK BASED APPROACH TO PRE-STARTUP SAFETY REVIEW

Item	Responsibility	Projected Completion Date	Completion Date	Initials
Water Circulating Systems Operational				
Nitrogen System Operational				
Areas Evaluated for High Noise Levels				
Regulated Areas/Confined Spaces Marked in accordance with OSHA				
Structural Steel/Operating Equipment Grounded				
Power Driven Equipment Guarded				
All Safety Signs In Place				
Hot Surfaces Insulated or Guarded				
CONTROL ROOM REQUIREMENTS:				
Operator Controls Operational				
Communications Operational				
Electrical Meets OSHA Code				
Temporary Electrical Removed				
Alarm Panels In Service				
Ventilation System Operational				
Fire Alarm System In Service				
Fire Suppression System In Service				
Fire Extinguishers In Place				
SCBA and/or Escape Packs In Place				
Protective Clothing In Place				
TV Monitors Operational				
Graphic Control Panel Operational				
Computer Control Systems Operational				
Process Safety Information In Place				
PROCESS CONTROL SYSTEMS:				
System Problems Cleared				

Item	Responsibility	Projected Completion Date	Completion Date	Initials
Alarms & Trips at Proper Settings				
Field Switch Alarms & Trips at Proper Settings				
Graphics Correct				
Required Loops Checked/Operational				
Gauges In Place & Operational				
Instrumentation Orientated for Easy Reading				
Control Valves Tested/Calibrated				
ENVIRONMENTAL:				
Environmental permits completed				
Review changes in project design scope to verify permit conditions will not be violated				
Verify equipment and piping components have been identified for tagging in the leak detection and repair program				
Verify process vents in hazardous air pollutant service are not vented to the atmosphere				
Verify closed loop sampling systems have been installed, if required.				
Verify waste production volumes have been evaluated and characterized.				
Attach any other process specific requirements developed by the PSSR team.				

4.5 TWO EXAMPLES OF USING A RISK-BASED APPROACH TO PSSR DESIGN

Here are two examples of using a risk-based approach in PSSR. The first uses the simple PSSR short form. The second example discusses using the PSSR long form.

4. A RISK BASED APPROACH TO PRE-STARTUP SAFETY REVIEW 51

Consider how these trigger events might be treated at other facilities. Are there special chemical or work practice considerations that may apply? Does the nature of the process come into play?

4.5.1 A Simple PSSR

Here is an example of applying the algorithm presented previously in Section 4.2.2 to a lower risk trigger event. We will use the risk matrix to determine the potential impact.

The trigger event for this example is the addition of a check valve to an existing bypass line.

1. Become aware of a trigger event (a change in regard to process safety or other organizationally defined standard).
   - The MOC activity is started if the trigger event is a change.

2. Assess the event's need for management of change and pre-startup safety review.
   - Change affects process safety information; therefore, it needs a PSSR.

3. Apply a qualitative risk-based analysis to the event to determine the event's potential severity and likelihood.
   - The example risk matrix in Table 4-1 will be used in this case.
   - There is a low probability of a catastrophic release for this line.
   - There is a low frequency of the check valve failing catastrophically.

4. Assess the complexity of the PSSR trigger event using the risk-based analysis results.
   - A simple PSSR is recommended for this event. *(NOTE: It is never incorrect to use the complex PSSR approach and complete the long form for any trigger event.)*

5. Assign the PSSR short form team.
   - The MOC initiator and the maintenance installation crew leader are assigned. The MOC initiator assumes the role of team leader.

6. Schedule document reviews and physical reviews.
   - The two team members meet to evaluate the status of the documentation. They determine the following:
     o No operating procedures need to be revised.
     o Safe work practices already exist both for the installation and the operation phase.

- o Maintenance procedures are adequate for this fairly standard trigger event.
- o The process overview will not be affected by this minor change.
- o The current emergency plan addresses any potential incidents.
- o However, the P&IDs must be changed to reflect the new check valve.
- The two team members determine at least one of them will inspect the installation upon completion and schedule the visit to the installation.

7. Follow up on action items.
   - As assigned, each follows the P&ID revision request and the physical verification of proper installation.

8. Document the equipment is safe for startup using the PSSR short form when all criteria are met.
   - When satisfied the two action items are complete and the management of change system has been applied properly, each signs the confirmation line of the PSSR short form. An example completed short form follows.

4. A RISK BASED APPROACH TO PRE-STARTUP SAFETY REVIEW 53

TABLE 4-4 Example completed short form PSSR

PRE-STARTUP SAFETY REVIEW – SHORT FORM			
Date: 3/1/2007	PSSR Team Leader: A. Engineer		
Facility/Process Equipment Reviewed: Installation of check valve on product bypass line. MOC ID No. 12345			
Type of Startup (Check one)	New Construction: ☐	Modified Process: ☑	
Recommendations: (Expand as needed. Attach relevant documents to this form)			
Item No.	Description	Initials	Date Resolved
12345-1	Update P&ID to reflect change	A.E.	3/15/2007
12345-2	Verify installation	A.E.	3/19/2007
PSSR Completion Summary: The following issues have been resolved and the undersigned believe the process/facility is ready for startup.			
• The construction and equipment meet design specifications.			
• Safety, operating, maintenance, and emergency procedures are in place and are adequate.			
• For new facilities, the initial process hazard analysis (PHA) has been performed and recommendations have been resolved.			
• Training of each employee involved in the operating process is complete.			
• Changes made to modify the process/facility have been reviewed and authorized by the facility management of change program.			
Confirmation by PSSR Team Members			
Name/Title	Signature	Date	
A. Engineer – Process Engineer	/s/ A.E.	3/20/2007	
D. Millwright – Outside Maintenance Crew Leader	/s/ D.M.	3/20/2007	

4.5.2 A More Complex PSSR

Here is an example of applying the algorithm to a higher risk trigger event. We will use the Table 4-1 – *Typical Risk-based Decision-making Steps in a PSM PSSR Procedure* to determine the potential impact.

The trigger event for this example is the modification of an existing process to include a new chemical additive injection system. The major components to consider for this trigger event consist of:

- A 500-gallon permanent chemical storage tank with dike and access scaffold
- Up to 500-gallons of the proprietary additive chemical (health rating of 3)

- Filling and draining pumps for tank service and refill
- Level instrumentation for the tank
- Metering pumps and injection tubing
- In-line analyzer with signal feed to metering pump
- Project cost is estimated at $20,000

1. Become aware of a triggering event (a change in regard to process safety or other organizationally defined standard).
   - The MOC activity is started if the trigger event is a change.

2. If the change management review indicates the process safety information must be changed, ask the following questions about the change to determine if it is classified as a simple or complex PSSR. When any one of the following questions can be answered as *yes*, use the PSSR Long Form to plan and implement the review:

   YES ☑ NO ☐   The change involves equipment in a process that involves materials with:
   - health ratings of 3 or 4,
   - reactivity ratings of 3 or 4, or
   - flammability ratings of 3 with operating temperatures above 75 degrees F or operating pressures above 15.2 psig, and
   - all materials with flammability rating of 4.

   YES ☑ NO ☐   The project cost is greater than $10,000.

   YES ☐ NO ☑   The change involves a new type of equipment (that is, the first of its kind at the site or substantially different from other equipment on site).

   YES ☐ NO ☑   The change involves three (3) or more tie-in points.

   YES ☑ NO ☐   If the process involved in the change fails; it is likely to result in a reportable incident for the facility.

   YES ☐ NO ☑   The change involves new control systems or modifications that affect safety controls or interlocks (both safety & non-safety)

4. A RISK BASED APPROACH TO PRE-STARTUP SAFETY REVIEW

      YES ☐ NO ☑   The change involves new or modified fire protection systems.

For our example, the project cost, environmental concerns, and the health rating of the new chemical additive guide us to use the PSSR long form.

3. Assemble the PSSR long-form team.
   - The MOC initiator assumes the role of team leader for this example.
   - Team size is determined based upon the technical aspects of the trigger event and the nature of the risk.
   - For our example, the team may need to include instrumentation specialists, industrial hygiene specialists, construction personnel, process operations, and engineering specialists for the process to which the chemical addition system is being added.

4. Work with the team to examine the PSSR long-form checklist (part 2 of the example form).

5. The PSSR leader should facilitate the team to identify those areas within their expertise and then determine whether specific items need special review.

6. Work with the project team's installation target date to schedule physical reviews, document preparation, training, and other activities as needed. Try to meet the intended schedule, but never sacrifice pre-startup safety review quality for a deadline.

7. This form shows all of the applicable items (those not marked as NA), which team member is responsible for an item's review, and the projected date for the review to be completed.

8. Schedule and document physical inspections for a long term project or one involving construction as in this example.

9. Prepare a project Gantt chart if needed to help track PSSR progress along with the installation project's progress.

TABLE 4-5 Example long form PSSR in progress

PART 1 - PRE-STARTUP SAFETY REVIEW - LONG FORM			
Date: 3/1/2007	PSSR Team Leader: A. Engineer		
Facility/Process Equipment Reviewed: Proprietary chemical addition system installation MOC ID No. 54321			
Type of Startup (Check one)	New Construction: ☐	Modified Process: ☑	
Recommendations: (Expand as needed. Attach relevant documents to this form)			
Item No.	Description	Initials	Date Resolved
54321-1	See Form and Gantt chart	See Form	OPEN
through			
54321-81	See Form and Gantt chart	See Form	OPEN
PSSR Completion Summary: The following issues have been resolved and the undersigned believe the process/facility is ready for startup.			
• The construction and equipment meet design specifications.			
• Safety, operating, maintenance, and emergency procedures are in place and are adequate.			
• For new facilities, the initial process hazard analysis (PHA) has been performed and recommendations have been resolved.			
• Training of each employee involved in the operating process is complete.			
• Changes made to modify the process/facility have been reviewed and authorized by the facility management of change program.			
Confirmation by PSSR Team Members			
Name/Title	Signature		Date
A. Engineer / Area Engineer			
A. Instrument / I&E Tech			
D. Millwright / Mechanical Integrity Dept.			
D. Trainer / HR Dept.			
A. Safety /EHS/IH Dept.			

4. A RISK BASED APPROACH TO PRE-STARTUP SAFETY REVIEW

Item	Responsibility	Projected Completion Date	Completion Date	Initials
PROCESS HAZARD ANALYSIS ISSUES:				
Pre-startup action items closed	D.T.	3/15/07		
Post-startup action items ID'd	NA			
Other:	NA			
DOCUMENTS IN PLACE:				
Operating Procedures	D.T.	3/15/07		
Maintenance Procedures	D.T.	3/15/07		
Safety Procedures	A.S.	3/15/07		
Emergency Response Procedures	NA			
Other: Process Overview	D.T.	3/15/07		
PROCESS SAFETY INFORMATION:				
P&IDs	A.E.	3/15/07		
Process Flow Diagrams	A.E	3/15/07		
Safety Equipment Plot Plan	NA			
Area Electrical Classification	NA			
MSDS/Chemical hazards	A.S.	3/15/07		
Safe Operating Limits	A.E.	3/15/07		
Process Description/Chemistry	A.E.	3/15/07		
Equipment Lists	D.M.	3/15/07		
Ventilation Systems	NA			
Heat & Material Balance	A.E.	3/15/07		
Relief Valve Summary List	A.E.	3/15/07		
Instrument Safety System List	A.E.	3/15/07		
Mechanical Safety System List	A.E.	3/15/07		
Piping Schedule	A.E.	3/15/07		
Maximum Intended Inventory	A.E./A.S.	3/15/07		
MECHANICAL INTEGRITY INFORMATION:				
Rotating Equipment Records	D.M.	3/25/07		
Electrical Equipment Records	D.M.	3/25/07		
Inspection Records	D.M.	3/25/07		
Relief Valve Records	D.M.	3/25/07		
Equipment Records	D.M.	3/25/07		
Equipment Specifications	D.M.	3/25/07		

Item	Responsibility	Projected Completion Date	Completion Date	Initials
Electrical 1 line drawings	D.M.	3/25/07		
INSTRUMENTATION & CONTROLS INFORMATION:				
Advanced Controls Documentation	NA			
DCS or PLC Documentation	A.I.	3/25/07		
DCS Screen Revisions	A.I.	3/25/07		
TRAINING: (for startup personnel at a minimum)				
Operator/Maintenance Process Overview Training	D.T.	4/13/07		
Operator Procedure Training	D.T.	4/13/07		
Emergency Shutdown Procedures	NA			
Emergency Evacuation Procedures	NA			
Maintenance Procedure Training	D.T.	4/13/07		
HAZCOM Training Certified	D.T.	4/13/07		
Fire Suppression System Training	NA			
Fire Department Training	D.T.	4/13/07		
Emergency Response/Hazmat Training:				
Level 1: Awareness	NA			
Level 2: Operations	NA			
Level 3: Technician	NA			
EPA Hazardous Waste Training	D.T.	4/13/07		
Safety Shower/Eye Wash Training	A.S.	4/13/07		
GENERAL FACILITY REQUIREMENTS				
Construction/Equipment Meet Design Specifications	A.E.	4/13/07		
Ladder/Platforms Meet OSHA Standards	A.S.	4/13/07		
Handrails/Toe boards/Walkways Meet OSHA Standards	A.S.	4/13/07		
Electrical Systems In Service	A.I.	4/13/07		
Lighting Adequate	A.S.	4/13/07		

4. A RISK BASED APPROACH TO PRE-STARTUP SAFETY REVIEW

Item	Responsibility	Projected Completion Date	Completion Date	Initials
Safe Access to Block Valves, gauge glasses, and Bleeders	A.S.	4/13/07		
Temporary Electrical Removed	NA			
Electrical meets OSHA Code	A.E.	4/13/07		
Fire Protection Systems Tested	NA			
Fire Alarm Systems Working/Tested	A.S.	4/13/07		
Fire Extinguishers In Place	A.S.	4/13/07		
Fire Hydrants & Monitors In Place	NA			
Fire Hoses In Place	NA			
Fire Water Pumps Operational	NA			
Oxygen/Gas Monitors/Alarms Calibrated	A.I.	4/13/07		
Utilities Marked in accordance with OSHA Standard	A.S.	4/13/07		
Identify Piping System Dead legs and Eliminate if Possible	A.E.	4/13/07		
Containers and Piping Marked in accordance with OSHA Standard	A.S.	4/13/07		
Safety Shower/Eye Wash Operational	A.S.	4/13/07		
Gas Rescue Equipment In Place	NA			
Steam Systems In Place/Operational	NA			
Control Loops Checked/Operational	A.I.	4/13/07		
Steam Tracer Circuits Active	NA			
Steam Lines Insulated in accordance with OSHA Standard	NA			
Steam Vents & Pressure Relief Valves Operational	NA			
Instrument Air Systems Operational	A.I.	4/13/07		
Instrument Air Dryer Systems Operational	NA			
Water Circulating Systems Operational	A.E.	4/13/07		

Item	Responsibility	Projected Completion Date	Completion Date	Initials
Nitrogen System Operational	A.E.	4/13/07		
Areas Evaluated for High Noise Levels	A.S.	4/13/07		
Regulated Areas/Confined Spaces Marked in accordance with OSHA	A.S.	4/13/07		
Structural Steel/Operating Equipment Grounded	A.E.	4/13/07		
Power Driven Equipment Guarded	A.S.	4/13/07		
All Safety Signs In Place	A.S.	4/13/07		
Hot Surfaces Insulated or Guarded	NA			
CONTROL ROOM REQUIREMENTS:				
Operator Controls Operational	A.E.	4/13/07		
Communications Operational	NA	4/13/07		
Electrical Meets OSHA Code	A.E.	4/13/07		
Temporary Electrical Removed	NA			
Alarm Panels In Service	A.E.	4/13/07		
Ventilation System Operational	NA			
Fire Alarm System In Service	A.E.	4/13/07		
Fire Suppression System In Service	NA			
Fire Extinguishers In Place	NA			
SCBA and/or Escape Packs In Place	NA			
Protective Clothing In Place	A.S.	4/13/07		
TV Monitors Operational	A.E.	4/13/07		
Graphic Control Panel Operational	A.E.	4/13/07		
Computer Control Systems Operational	A.E.	4/13/07		
Process Safety Information In Place	A.E.	4/13/07		
PROCESS CONTROL SYSTEMS:				
System Problems Cleared	A.I.	4/13/07		

4. A RISK BASED APPROACH TO PRE-STARTUP SAFETY REVIEW

Item	Responsibility	Projected Completion Date	Completion Date	Initials
Alarms & Trips at Proper Settings	A.I.	4/13/07		
Field Switch Alarms & Trips at Proper Settings	A.I.	4/13/07		
Graphics Correct	A.I.	4/13/07		
Required Loops Checked/Operational	A.I.	4/13/07		
Gauges In Place & Operational	A.I.	4/13/07		
Instrumentation Orientated for Easy Reading	A.I.	4/13/07		
Control Valves Tested/Calibrated	A.I.	4/13/07		
ENVIRONMENTAL:				
Environmental permits completed	NA			
Review changes in project design scope to verify permit conditions will not be violated	NA			
Verify equipment and piping components have been identified for tagging in the leak detection and repair program	A.S.	4/13/07		
Verify process vents in hazardous air pollutant service are not vented to the atmosphere	A.E.	4/13/07		
Verify closed loop sampling systems have been installed, if required.	NA			
Verify waste production volumes have been evaluated and characterized.	NA			
Attach any other process specific requirements developed by the PSSR team.				

Chapter 6 – *Methodologies for Compiling and Using Your PSSR Checklist* describes electronic options for expanding or collapsing the PSSR long-form checklist.

4.6 REFERENCES

4-1. Occupational Safety and Health Administration, *Process Safety Management of Highly Hazardous Chemicals*, 29 CFR Part 1910, Section 119, Washington, DC, 1992.

4-2. Environmental Protection Agency, *Accidental Release Prevention Requirements: Risk Management Programs*, Clean Air Act, Section 112 (r)(7), Washington, DC, 1996.

4-3. American Institute of Chemical Engineers, *Guidelines for Implementing Process Safety Management Systems*, Center for Chemical Process Safety, New York, NY, 1994.

4-4. T. A. Kletz, Learning from Accidents, 2nd edition, Butterworth-Heinemann, Oxford, 1994.

5
THE PRE-STARTUP SAFETY REVIEW WORK PROCESS

In order to manage a process safety and environmental risk program element like pre-startup safety review (PSSR), it is essential to understand the work process. As we have discussed, different facilities have different levels of risk, levels of resources, and organizational cultures to consider. If covered under OSHA PSM or EPA RMP, each facility must be able to demonstrate compliance, but how it is specifically accomplished is open to interpretation.

This chapter addresses the basics of building a pre-startup safety review program and items to consider for building or upgrading a PSSR management system. The following topics are included:
- Defining the PSSR system
- Understanding the four PSSR sub-elements
- Designing an initial PSSR program
- Preparing for performing a PSSR
- Following pre-startup safety review action items
- Using the PSSR forms, checklists, or databases
- Validating PSSR is complete
- Approving startup in accordance with PSSR

5.1 DEFINING THE PSSR SYSTEM

Any process safety management or environmental risk management program element needs to be defined in the administrative document which describes the activities that make up the element for a facility. This should be a good goal for someone upgrading the PSSR program for a facility. Think about it; every document that has any value to an organization describes an activity – every form, every procedure, every policy. They all imply action. So defining the PSSR system within a larger system is essential, and a clear definition of your intent is helpful to all users. Here is an example list of typical administrative procedures that make up the process safety and environmental risk management documents for an OSHA

PSM/EPA RMP covered facility. It establishes conventions used in the example PSSR administrative procedure offered later in this book.

TABLE 5-1 An example integrated PSM/RMP compliance plan and management system

Document Control Number	Typical Document Title
PSM-ADM-01	PSM / RMP Compliance Plan
PSM-ADM-02	Employee Participation
PSM-ADM-03	Process Safety Information
PSM-ADM-04	Process Hazard Analysis
PSM-ADM-05	Procedure Program
PSM-ADM-06	Training Program
PSM-ADM-07	Contractors
PSM-ADM-08	Pre-Startup Safety Review
PSM-ADM-09	Mechanical Integrity
PSM-ADM-10	Hot Work Permit Program
PSM-ADM-11	Management Of Change
PSM-ADM-12	Incident Investigation
PSM-ADM-13	Emergency Planning And Response
PSM-ADM-14	Compliance Audits
PSM-ADM-15	Trade Secrets
PSM-ADM-16	RMP Management System
PSM-ADM-17	RMPlan Executive Summary

Building an integrated management system for safety, environmental quality, and business excellence demands a rigorous approach to knowing your requirements, knowing your processes, and documenting how you operate and maintain performance within those boundaries. The example system here could easily make up part of a facility's overall business management system.

5.1.1 Double Checking Management of Change

In the case of management of change, the pre-startup safety review activity is focused on a trigger event that is a change – something not in the same condition it

5. THE PSSR WORK PROCESS

was before the trigger event – whether it is defined by law or by your facility's designated needs.

In these cases, the role of the PSSR program is to provide a second layer of protection around the management of change element. Some companies even bundle these elements together in one MOC/PSSR administrative-level management system procedure.

Essentially, in terms of applicable regulatory requirements, PSSR is verifying the four PSM/RMP requirements, which is essentially that the MOC process was followed. How can we make sure we have addressed these requirements in a way that adds value to the process? By using PSSR thoroughly and systematically, even when the benefit may not be explicit.

For non-change type trigger events, PSSR is verifying that the process is safe for startup, all maintenance has be completed, and other facility requirements have been met (for example, a PSSR following a maintenance turnaround or after an emergency shutdown). Again, the PSSR provides a second layer of protection that all other policies and procedures are complete.

5.1.2 Who Is Responsible for Driving the System?

Identifying the roles that personnel must fill to perform any PSSR, whether a simple or a complex PSSR, is essential. People like to know exactly what the requirements are for their roles in any task.

However you define your PSSR program, it should provide detailed responsibility and authority levels. The individuals who fill these roles should be trained in how they interface with the PSSR element. A charter should be clear for the following roles:

- The PSSR team leader is responsible for driving the PSSR process and may be responsible for discrete review tasks.
- The PSSR team is responsible for team member assignment to discrete review items and for confirming the pre-startup safety review is complete and the equipment is safe to operate.
- The MOC initiator/coordinator is responsible for verifying PSSR is complete before pursuing authorization to startup the change (if the example is a change based PSSR).
- The manager who owns the process or equipment involved in the PSSR has ultimate responsibility of authorizing any PSSR-related startup, typically through the interacting management of change element. For some facilities, multiple approvals may be necessary by several different persons. For example, engineering, production, and safety representatives approve startup of all PSM required MOC at some plants.

5.2 PSSR SUB-ELEMENTS

As we discussed briefly in Chapter 4, if a facility is affected by OSHA PSM or EPA RMP, there are very specific requirements that apply. Otherwise, a facility may follow local regulations and define its own requirements.

The common feature among the four regulated sub-elements is that each must be verified for each trigger event. Let's visit these again briefly.

5.2.1 Construction and equipment meet the designed specifications.

Design specifications for construction and equipment can be validated by:
- obtaining field verification,
- performing document review for the new or modified process, and
- if a change is not physical (such as a set point for an interlock shutdown), the method for the change and its anticipated effects shall be reviewed.

5.2.2 Safety, operating, maintenance and emergency procedures are in place and adequate.

Entries in the management of change authorization package or electronic database should indicate any operating, maintenance or emergency procedures that were developed or revised for the modification.

The MOC data package and referenced documents should be reviewed to determine if these documents are to be changed and to check on progress. Existing site safety procedures should be considered to ensure they exist and are adequate.

5.2.3 A PHA has been performed for new facilities.

The MOC data packages and referenced documents indicate when a process hazard analysis (PHA) in accordance with OSHA PSM/EPA RMP has been performed for the modification or new facility.

The PSSR team should verify that all of the pre-startup required PHA recommendations have been implemented or resolved before the facility can be judged safe to operate.

5.2.4 Training of each employee involved in the process is complete.

A completed section within the MOC data package should indicate when training of each employee involved in the startup of a new or modified process is complete.

Training of employees not directly involved with initial startup is also an issue of concern. When blended properly with a facility's MOC data package requirements, the PSSR team may consider training complete for introduction of highly hazardous chemicals into the new facility or modification when the MOC data indicates it is complete. If this is not possible, the PSSR team may need to go directly to the training records for the individuals to evaluate the status of training completeness. There should also be a mechanism to identify and train those employees who were on leave when the trigger event occurred.

5.2.5 General requirements

A completed section within the MOC data package should indicate when training of each employee involved in the startup of a new or modified process is complete.

5.3 DESIGNING AND IMPLEMENTING AN INITIAL PSSR PROGRAM

Facilities building a pre-startup safety review program from scratch or upgrading their program extensively should consider the following items.

5.3.1 Defining a Policy on PSSR

The PSSR administrative procedure or other process safety management system document can state the organization's policy in relatively straightforward terms.

The following is one example of such a policy statement. Modify it as needed to match your company's basic approach to pre-startup safety review.

> *A pre-startup safety review will be performed before the startup of a new or significantly modified facility is authorized.*
> - *A pre-startup safety review is required if the modifications to a facility are significant enough to require a change in the process safety information or*
> - *if it is a trigger event listed in this procedure.*
>
> *Risk ranking will be performed for every trigger to determine the nature and scope of the PSSR.*

This policy statement states the organization's commitment to PSSR, when it is required (but not limiting its use), ties it to the management of change element and process safety information element of the OSHA process safety management regulation and the EPA risk management program rule, and includes those cases where a process hazard analysis is required. Remember, some facilities establish policies for performing PSSR on trigger events unrelated to changes in the process safety information.

5.3.2 Defining the PSSR Team

A PSSR team will be formed as a part of new facility construction and when any trigger event modifies the process safety information for a process. Many companies require that the team will have at least one member besides the PSSR team leader. However, some companies do allow for one person to perform the PSSR for lower-risk (simple) PSSRs.

The coordinator for the modification requiring management of change or new facility typically serves as or designates the PSSR team leader. This person is often a facility engineer.

When using the PSSR long form for a complex PSSR, consider the following categories of personnel for members of the PSSR team:

- a process engineer
- a corporate or regional engineer
- a process control engineer with safety systems experience
- an area or process supervisory representative (depending upon your site's organization)
- operations personnel with appropriate knowledge and skills
- a mechanical representative and
- a safety representative

The PSSR team leader's primary function is to ensure the appropriate personnel and expertise is available on the team to review new facilities and major modifications thoroughly.

5.3.3 Designing the Specific PSSR

Here is where the risk-based approach necessitates a first look at the nature of the trigger event. By following some of the guidelines in Chapter 4, the type of PSSR to design is relatively easily divided into simple or complex – that is, a short form or a long form respectively. However, when the PSSR long form is indicated, there are numerous decisions to make about the pre-startup safety review's needs.

1. The first step is to use the long-form's checklist or your electronic PSSR database of questions to identify which are applicable and which are not.
2. Next look at the applicable items and determine whether they are:
   - Repetitive through the course of the trigger event's installation (that is, do we need to look at the item repeatedly or do we just check it once)
   - Something that occurs early in the trigger event's review or something that occurs at the end

5. THE PSSR WORK PROCESS

3. Schedule the activities as indicated. Again, some PSSRs may require discrete review activities that occur over a long period. Simpler PSSRs may only require one review.
4. Record the status of the review activities as they are performed.
5. Assign responsible persons and estimated due dates for follow up activities arising from the review activities.

The methods by which these design steps are scheduled, documented, and tracked are typically described in the PSSR's administrative procedure.

5.3.4 Training the Workforce on the PSSR Program

A facility's PSSR administrative document plus the text of any regulations which apply at the site become the source of the training programs discussed in Chapter 2 – *What is a Pre-startup Safety Review*?

These reference documents allow a training group or the site-training specialist to design modules for the audiences identified previously. Those are:

- PSSR team leaders
- PSSR team members
- Facility management personnel
- The remaining workforce

Each PSSR training module should be focused on objectives designed for the specific tasks each category of trainee is expected to perform or for transferring the knowledge they are expected to have regarding PSSR. Workers at any level should not be expected to perform a PSSR task unless they have been trained on it in some way – whether the training occurs in a classroom, on a computer-based training module, through directed self study, or through one-on-one on-the-job training. It is always a good practice to document this training in the employee's training records.

5.3.5 An Example PSSR Program

The example PSSR administrative procedure below shows one way to record the philosophy, methods, and documentation of adherence to the methods this book describes.

Compare this example to an existing PSSR program. Are there areas where one is more conservative than the other from a safety standpoint? Does one require tighter management control than the other? Consider the culture and resources available and identify aspects which could be addressed in a way that may enhance compliance or display the organization's philosophy toward PSSR and PSM in general. Examples of the associated PSSR short form and PSSR long form referenced in this example are provided in Chapter 4 – *A Risk Based Approach to PSSR*.

TABLE 5-2 Example pre-startup safety review administrative procedure

PSM-ADM-08 Rev.0	Page x of y
Anychem Co. – Anytown Facility	PSM/RMP Administrative Procedure
Pre-startup Safety Review	
PSSR Policy	A pre-startup safety review will be performed before the startup of a new or significantly modified facility is authorized. How to determine when a PSSR is required follows: • A pre-startup safety review **IS** required if the modifications to a facility are significant enough to require a change in the process safety information. (PSM required) • A pre-startup safety review **IS NOT** required for facilities that have been modified so slightly that process safety information does not change. (PSM required)
MOC Connection	However, for all modified facilities, management of change (MOC) requirements must be satisfied before startup. (PSM required)
New Facilities	For new facilities, a process hazard analysis must be performed before startup. The process hazard analysis recommendations must be resolved or implemented prior to startup. (PSM required)
Form Selection	A pre-startup safety review short form or long form will be signed to indicate completion of the review and resolution of any issues in accordance with Anychem Co. policy.
PSSR Team Description in accordance with Anychem Co. Policy	The PSSR team will be formed as a part of new facility construction or when any change modifies the process safety information for a process. The team will have at least one member besides the PSSR team leader. The MOC coordinator for the modification or new facility will serve as (or designate) the PSSR team leader. This person is normally a process engineer. When using the PSSR long form, the PSSR team will normally consist of a facility engineer, an area superintendent, and operations personnel with appropriate knowledge and skills. A mechanical representative and safety representative may also be included on the team. The PSSR team leader should ensure the personnel and expertise is available for the team to review new facilities and major modifications thoroughly as an Anychem Co. best practice.

5. THE PSSR WORK PROCESS

PSM-ADM-08 Rev.0	Page x of y
Anychem Co. – Anytown Facility	PSM/RMP Administrative Procedure

Pre-startup Safety Review

Release of Final Authority in accordance with Anychem Co. Policy

If a PSSR is **NOT** required, the area manger's approval signature on the MOC form is final authority that all requirements from previous sections of the MOC form are completed to his or her satisfaction and startup may occur.

If a PSSR **IS** required, the area manager's signature on the MOC form is STILL final authority that all management of change requirements are met and startup may occur. Since the MOC process indicates a PSSR is required due to changes in the process safety information, we enforce a check on that final authority.

The PSSR team's review and signatures are a mechanism to release the area manager's final authority after the PSSR team has double-checked specific requirements and, in the case of physical changes, obtained field verification the work was done.

PSSR Form Usage in accordance with Anychem Co. Policy

The PSSR team leader will evaluate the extent of the modification or new facility and determine which of the PSSR forms to use. These forms are described below:

- **PSSR Short Form** – This form is to be used for simple modifications where a PSSR is required due to minor changes in process safety information. A guideline for when to use this form is if a MOC for the modification's hazard review was able to be completed without using a process hazard analysis.

- **PSSR Long Form** – This form is to be used for new units or major process modifications. The PSSR team will establish specific target areas for the new facility or modification project and define the expected state of readiness for each target. The PSSR team may modify and customize the form or develop and attach additional items specific to the process under consideration.

When using either the long or the short form, the PSSR team will confirm that the following PSM/RMP requirements have been met before highly hazardous chemicals or energy sources are introduced into the new or modified facility.

PSM-ADM-08 Rev.0	
Anychem Co. – Anytown Facility	PSM/RMP Administrative Procedure

Pre-startup Safety Review

- **Construction and equipment meet the designed specifications.**

In accordance with Anychem Co. policy, design specifications for construction and equipment shall be validated by obtaining field verification and performing document review for the new or modified process. If a change is not physical (such as a set point for an interlock shutdown), the method for the change and its anticipated effects shall be reviewed.

- **Safety, operating, maintenance and emergency procedures are in place and adequate.**

In accordance with Anychem Co. policy, entries in the MOC form will indicate any operating, maintenance or emergency procedures that were developed or revised for a modification. The MOC package and referenced documents must be reviewed. Existing site safety procedures must be checked to ensure they exist and are adequate.

- **A PHA has been performed for new facilities.**

In accordance with Anychem Co. policy, the MOC packages and referenced documents indicate when a process hazard analysis (in accordance with PSM-ADM-04 - *Process Hazard Analysis*) has been performed for the modification or new facility. The PSSR team must verify that all of the PHA recommendations required before startup have been implemented or resolved before the facility can be judged safe to operate.

- **Training of each employee involved in the operating process is complete.**

In accordance with Anychem Co. policy, a completed MOC form indicates when training of each employee involved in the startup of a new or modified process is complete. Training of employees not directly involved with startup is described in PSM-ADM-06 - *PSM Training Program*. When the MOC is signed, the PSSR team may consider training complete for introduction of highly hazardous chemicals or energy into the new facility or modification.

5. THE PSSR WORK PROCESS

PSM-ADM-08 Rev.0	Page x of y
Anychem Co. – Anytown Facility	PSM/RMP Administrative Procedure

Pre-startup Safety Review

Change Management	In accordance with Anychem Co. policy, an overall review of the MOC package must be performed for modified facilities to ensure all: • priority 1 – (required before startup), and • priority 2 – (allowed after startup) update items (including material safety data sheets) have been addressed and that priority 1 items are complete.
Conduct the PSSR	In accordance with Anychem Co. policy, the PSSR team leader will: • Schedule and conduct PSSR team meetings and field inspections as required by the complexity of the change • Verify that all applicable sections of the selected PSSR form have been considered • Identify issues which **MUST** be corrected **BEFORE** startup and issues which **WILL** be corrected **AFTER** startup. In accordance with Anychem Co. policy, decisions for categorizing the issues for PSM compliance should be based upon the following: **BEFORE**- A deficiency that could cause, or result in, actual or potential release of hazardous chemicals to environment. The process cannot be safely started or operated until the issue is corrected. - A priority 1 PSM/RMP requirement from the MOC package is not satisfied. - An apparent unsafe condition exists **AFTER** - An issue that does not impact safe startup or operation but, if corrected, enhances process safety. - A priority 2 PSM/RMP requirement from the MOC package is not satisfied. - Process safety information needs to be permanently updated (for example, piping and instrumentation diagrams are marked up and logged for change but not necessarily reissued).

PSM-ADM-08 Rev.0	Page x of y
Anychem Co. – Anytown Facility	PSM/RMP Administrative Procedure

Pre-startup Safety Review

Progress Audits	In accordance with Anychem Co. policy, for extensive modification or construction, periodic PSSR audits will be conducted to measure of the progress in each target area to achieve the expected state of readiness goals. The PSSR team will establish a timetable to ensure the periodic audit schedule is completed and that recommendations are published to effect corrective action. The PSSR long form may be used as a guide to perform the periodic audits and track the completion of the critical items.
Completing the PSSR Form	In accordance with Anychem Co. policy, the PSSR team will perform a physical review of the facility just before startup to confirm that all related requirements have been met before highly hazardous chemicals are introduced. All members of the PSSR team will review and sign the PSSR form to confirm the facility is safe for startup. When using the PSSR long form, the PSSR team leader will assign responsibility for each item. Any items that do not apply to the specific review should be marked *not applicable* in the responsibility box.
PSSR Short Form Issues & Recommendations	In accordance with Anychem Co. policy, on the PSSR short form, a date in the *date resolved* box and initials in the *initials* box indicates that an item is satisfactory or that the issue has been corrected or resolved or a recommendation is closed.
PSSR Long Form Issues & Recommendations	In accordance with Anychem Co. policy, a date in the *completion date* box and initials in the Initials box adjacent to an item on the PSSR long form indicates that an item is satisfactory or that the issue has been corrected or resolved. A follow-up inspection will be performed if required. The first page has a space for special recommendations and, as with the short form, a date in the *date resolved* box and initials in the *initials* box indicates that an item is satisfactory or that the issue has been corrected or resolved or a recommendation is closed.
Closure In Accordance with Anychem Co. Policy	The responsible PSSR team leader will verify that all recommendations and issues have been satisfactorily resolved.

5. THE PSSR WORK PROCESS 75

PSM-ADM-08 Rev.0	Page x of y
Anychem Co. – Anytown Facility	PSM/RMP Administrative Procedure

Pre-startup Safety Review

The completed PSSR form will be sent to the PSM administrator and maintained as a part of a MOC project closure file.

END

5.4 PREPARING TO PERFORM A PRE-STATUP SAFETY REVIEW

The example PSSR program administrative procedure above provides the basic information to trigger the key activities a PSSR team needs to consider. The following sections identify some considerations that may need customization.

5.4.1 Gather the Documentation

Every facility has its own approach to maintaining plant information that may help a PSSR team complete its assigned tasks. And every PSSR may have slightly different needs in regard to the information necessary for a thorough review.

For some simple PSSRs, the team may only need access to the work order for a small change. For more complex PSSRs, the team may need to see P&IDs, schematics, MSDS, operating procedures, maintenance procedures, and safe work practices.

The advent of electronic document management systems, business management systems, training records management systems, and database-driven maintenance planning and general action item tracking systems provide powerful tools for those facilities where they have been implemented. Some facilities still use a mix of hardcopy and electronic access to process safety information and plant status. Others still use hardcopy approaches to their documentation across the plant. In any case, gathering these documents and determining if tasks are complete, document update was required, and if the update was effective is an essential part of preparing for and completing a PSSR.

5.4.2 Schedule Meetings as Needed

As previously described, the category of PSSR is the deciding factor for the number of meetings associated with the review. For some simple PSSRs, one trip to the physical location of a newly installed piece of equipment and a short document review may be all the evidence the team needs to satisfy the PSSR administrative procedure and sign the PSSR form. More complex PSSRs may need numerous meetings over the course of the modification period.

If your facility uses common group scheduling software, it can be relatively easy for the team to put regular meetings on their calendars and know who is available for the various meetings that may be needed.

Close and frequent contact with the project team leader and access to their work schedule for a complex trigger event will help the PSSR team leader in determining at least a window of time during which the various PSSR team meetings will be needed. The trigger event's major milestones are usually a good starting point for evaluating whether a PSSR team can effectively make physical or documentation inspections.

In plants with fewer personnel resources, some PSSR team members may very well be the same people participating in the project team for the trigger event being reviewed.

5.4.3 Verify the Trigger Event Related Work Is Complete

The PSSR team's job is easiest when the trigger event involves a straightforward physical installation or modification to a system or piece of equipment. When programming for safety instrumented systems (SIS) or other process control needs are involved, the PSSR team may need to make a special effort to verify the change is complete.

For physical trigger events, a team member or multiple team members can schedule a visit to the work location during construction or installation and again when the work is logged as complete. If the members have the proper expertise, they can quickly determine what has been done and what has yet to be done.

For software-related trigger events, the team may require a specialist in the programming arena to design a way to show that the change has been made and that it will operate as intended.

In either case, each team member must be satisfied that the trigger event has been addressed in a way that is safe for startup prior to putting their identification or signature (whether hardcopy or electronic) on the PSSR indicating the review is complete and all required activities have occurred or are identified.

5.4.4 Identify and Track the Process Hazard Analysis Action Items

If the trigger event, especially one that is covered by a process safety management or environmental risk management program management of change requirement, requires a process hazard analysis, the PSSR team has a task of verifying the PHA action items are resolved prior to authorizing startup.

It is common for a manager with ownership of the equipment involved with the trigger event to be accountable for resolving action items in their areas in a

5. THE PSSR WORK PROCESS

timely fashion. Most of the follow-up is done by the engineers and operations personnel. However, any employee may be assigned to follow-up.

An action item, including those for operability, can be resolved by:
- Performing the recommended action exactly as stated.
- Meeting the goal of the recommendation in a different way if further study shows the hazard can be controlled better or less expensively.
- Using a temporary solution while engineering is completed or while waiting to install a permanent solution. For instance, a valve might be locked open to protect a line from being blocked in until the next shutdown at which point the piping system would be upgraded; however, the action item is not closed until after the upgrade is complete.

All PHA action item resolutions must be documented and justification must be given if it is resolved differently from the method recommended. When a facility's PHA program is working properly, the PSSR team can easily check the status of the outstanding PHA action items.

If a facility is covered under EPA's risk management program (RMP) rule, changes in processes, quantities stored or handled, or any other aspect might reasonably be expected to increase or decrease the distance to the EPA RMP offsite consequence analysis endpoints by a factor of two or more, the facility risk management plan should be updated.

5.5 FOLLOW PRE-STARTUP SAFETY REVIEW ACTION ITEMS

As described above, for extensive modification or construction, periodic PSSR meetings will typically be conducted to measure the progress in each target area to achieve the expected state of readiness goals. The PSSR team will establish timetables to ensure the periodic audits are completed and that resulting recommendations are communicated for resolution. The following discussions use the example long and short forms from Chapter 4, Tables 4-2 and 4-3. Refer to them for ease of understanding. Remember, many aspects of how the PSSR work process is implemented are company policy and may not be an explicit requirement of the PSM or RMP regulations.

The PSSR long form may be used as a guide to perform the periodic reviews and track the completion of the critical applicable items.

The PSSR team should perform a physical review of the facility just before startup to confirm that all requirements have been met. Often, all members of the PSSR team review and sign the PSSR form to confirm the facility is safe for startup. When using the PSSR long form for complex trigger events, the PSSR team leader will have assigned responsibility for each applicable item.

On the PSSR short form for simple trigger events, a date in the *date resolved* box and initials in the *initials* box indicates that an item is satisfactory or that the issue has been corrected or resolved or a recommendation is closed.

A date in the *completion date* box and initials in the *initials* box adjacent to an item on the PSSR long form indicates that an item is satisfactory or that the issue has been corrected or resolved. A follow-up inspection will be performed if required. The first page has a space for special recommendations and, like the short form, a date in the *date resolved* box and initials in the *initials* box indicates that an item is satisfactory or that the issue has been corrected, resolved, or a recommendation is closed.

The PSSR team leader generally ensures that all recommendations and issues have been satisfactorily resolved.

5.5.1 Which Items Are Critical for Safe Operation?

Most facilities identify the critical items which must be in place prior to starting up the process. This makes it relatively straightforward for the PSSR team to identify what they must see or verify prior to completing the PSSR. However, trigger events a company identifies for PSSR but not tied to MOC may rely on the PSSR team's expertise to determine if any actions may be left for completion after the event has been reviewed and the process has been started.

All items critical for safe operation or required by regulatory requirements to be in place before introducing hazardous chemicals or energy into the equipment in question should be checked for installation, completeness, or compliance.

The PSSR team leader has the responsibility of ensuring adequate engineering, safety or environmental expertise is called upon when needed. It is the responsibility of the facility's leadership team to ensure the PSSR team gets the access they need to complete the review. The team leader is typically responsible for seeing the entire PSSR through to closure.

5.5.2 Consider Past PSSR PSM Compliance Audit Findings

It is a good practice to review past PSSR element PSM compliance audits for findings, their resolutions, and best practices (if your audit system calls these out). The PSSR team leader can review these and consider the trigger event in question to share applicable items with the team.

This practice makes use of the organization's knowledge base and avoids reinventing the wheel each time an event needs a PSSR.

5.6 APPROVE THE PRE-STARTUP SAFETY REVIEW REPORT

Some companies require PSSR team signatures as a way of encouraging ownership of the PSSR authorization to start up the equipment involved in the trigger event, whether a simple item or a complex modification. In any case, this is typically the last step in the PSSR work process.

5. THE PSSR WORK PROCESS

5.6.1 Reference the Documentation: Electronic or Hardcopy

Consider listing all the documentation whether electronic systems or hardcopy forms and reports used to conclude that the PSSR is completed. This can provide a valuable resource in the event of an incident and it is very helpful to future PSSR teams reviewing similar trigger events.

5.6.2 PSSR Team Approval

The PSSR form can be circulated in hardcopy for the team to sign if physical signatures are needed, but often the final team meeting provides a good venue for everyone on the team to reach consensus and ask any last minute questions they may have.

When using an electronic system, a tickler email may arrive asking the members to go to the software program for the PSSR in question and sign electronically or enter their name or other identification indicating their finding that the trigger event is ready for operation.

5.6.3 Management Approval

Some facilities have the completed PSSR package go to the manager who owns the equipment under review for review or actual approval. Some MOC programs use the PSSR completion as the last step of the management of change process. In any case, consider having the manager on the distribution so that he or she may be informed of the PSSR completion.

5.7 REFERENCES

5-1. Occupational Safety and Health Administration, *Process Safety Management of Highly Hazardous Chemicals*, 29 CFR Part 1910, Section 119, Washington, DC, 1992.

5-2. American Institute of Chemical Engineers, *Guidelines for Implementing Process Safety Management Systems*, Center for Chemical Process Safety, New York, NY, 1994.

5-3. Environmental Protection Agency, *Accidental Release Prevention Requirements: Risk Management Programs*, Clean Air Act, Section 112 (r)(7), Washington, DC, 1996.

6
METHODOLOGIES FOR COMPILING AND USING A PSSR CHECKLIST

An effective pre-startup safety review (PSSR) is centered on the tools a company provides to assist its staff in evaluating the safety and operational readiness of the trigger events it evaluates. One tool many companies provide is a PSSR checklist or electronic database of items to consider. This chapter examines options and considerations for compiling the items applicable to a facility.

6.1 BUILDING YOUR FACILITY'S DATABASE OF QUESTIONS

Depending upon the specific aspects of a facility's processes and the materials, intermediates, and products involved, the PSSR checklist questions should be customized to match those aspects. Consider selecting them in a way that guides the PSSR team to analyze the known hazards and associated risks for the total site. This chapter provides guidelines that may assist in building a database of standard questions to be considered for every complex PSSR.

6.1.1 Beware of Shortcuts

Adopting a PSSR checklist provided by an industry source, consultant, or even from another plant or corporate group within a site's own organization may seem efficient, and can be a good way to jump start the process, but it should still be subject to management, technical and EHS review against the needs of the site before it is adopted.

Of course there are some standard items or questions that can apply to any site, but a review against unique characteristics of the site when developing the collection of PSSR review items is a good practice. Review the example PSSR checklists in Appendix A. Are there differences? Are there similarities? Analyze the items in the examples for clues to the type of processes or materials present at the example checklists' site. Each represents a different approach to building a customized database of questions. None of the examples are intended to guarantee

6. METHODOLOGIES FOR COMPILING AND USING PSSR CHECKLISTS

perfection in the PSSR process. They are all simply tools to help diligent PSSR teams identify areas of interest for a wide range of trigger events.

6.1.2 Considerations for Different Industries

The term *chemical processing industry* (CPI) is wide ranging in its scope. Other manufacturers not included within the grouping CPI also can benefit from pre-startup safety reviews. The PSSR items used to spur thorough review for a refinery may be very different from those used for a pharmaceutical manufacturer, semiconductor manufacturer, or a coatings or specialty chemical manufacturing facility.

Look at the following types of issues when building, revising, or customizing your PSSR element's database of review items:

- Unique aspects of the process equipment
- Unique aspects of the technology
- Proprietary aspects of the process or units on site
- Characteristics of the raw materials or feedstock
- Characteristics of any intermediates involved
- Characteristics of the end products
- Reactivity issues between any of the chemicals, utility gases and fluids, or materials of construction used at the facility

6.2 VARIOUS APPROACHES: ELECTRONIC VERSUS HARDCOPY

Depending upon a site's resources and culture, the pre-startup safety review may involve a completely electronic approach to identifying database items which apply for a specific PSSR, a partial electronic approach in which the checklist database items are customized in a document using a word processing or spreadsheet software program prior to hardcopy distribution and use, or a completely paper driven hardcopy approach. All are suitable when the persons involved in the review understand the administrative, technical, and possible regulatory-driven aspects of the facility's PSSR element.

6.2.1 Using your Existing Facility Action Item Tracking System

Many companies have some type of electronic action item tracking system which can be used for all types of scheduled activities. This is a common tool in almost any business. It may be very simple or quite complex and customized in its operation and user interface. There may be several different action item tracking software applications available at a facility.

Whether an electronic or a hardcopy PSSR checklist approach is used at a site, each applicable item for a complex PSSR in progress can be entered into the action item tracking system and the following attributes can be addressed:

- The individual or department responsible for the review
- The recorded results of the review
- Any scheduled follow-up actions indicated or suggested
- The individual or department responsible for the following up on the actions
- The final closure of the item
- The actions associated with closure of the PSSR in total

6.2.2 Basic Electronic PSSR Checklist Tools

The most basic approach to an electronic PSSR checklist is to create a word processing software or spreadsheet software template form for the simple and complex checklists. If desired, the simple and complex items can be built together into one document as the instructions might be to either mark or delete all non-applicable items from the file customized for the specific PSSR trigger event.

Once the person responsible for the PSSR has customized the template checklist, either alone or by gaining team consensus, it can be printed out and transmitted as a hardcopy for the PSSR team to use and track progress, or it can be transmitted through electronic mail to the team members with specific instructions for their role in the review.

For more complex PSSRs occurring over a longer period of time, each team member may be responsible for tracking their own progress with regular checks by the PSSR team leader.

6.2.3 Electronic Change Management Systems with PSSR Tools

A common software tool some companies have adopted is an electronic management of change (MOC) system. There are numerous approaches to doing this. Some software is commercially available, other software is customizable for the client, and still other software has been developed in-house or in close collaboration with a custom software application development partner.

Many of these electronic MOC systems include PSSR as a segment of the MOC activity. In these cases, it is typical that when a PSSR is required or desired for a trigger event, a field is selected and the user is led through the work process for performing the PSSR and offered a database of PSSR checklist items to evaluate for applicability.

A benefit of these types of electronic MOC/PSSR tracking and authorization systems is that the completion of one step in the work flow is immediately logged and the process can, in some more advanced software applications, be moved along automatically to its next phase.

However, potential pitfalls may also be present. For example, the MOC initiator may miss marking that a PSSR is required and the management of change process may move forward blindly without realization that a key component of

6. METHODOLOGIES FOR COMPILING AND USING PSSR CHECKLISTS

managing the trigger event was missed. Even if the data field indicating a PSSR is needed gets selected, the PSSR database checklist items that apply may be missed and several other human factors oriented mistakes can be made. Another example of a potential electronic PSSR system pitfall is due to the removal of the human element. If the system automatically filters PSSR checklist items for the type of trigger event, potentially applicable items can be missed. But in a simple hardcopy system, a person may still choose to select sections of the form or specific items that at first glance may not seem applicable.

It should be obvious that these same pitfalls can occur with any hardcopy PSSR system; but computerized electronic systems help us make these mistakes much more efficiently.

6.3 AN EXAMPLE ELECTRONIC CHECKLIST

By modifying the basic example PSSR checklist from Chapter 4 as if it represented the database fields for an electronic checklist, the concept of expanding or collapsing the applicable items can be demonstrated.

6.3.1 Collapse the Checklist for Simple PSSR

Once the PSSR team has evaluated the risk-based aspects of the trigger event and determined a simple PSSR is suitable, the resulting software screens can be collapsed to show only those fields required.

An example of a possible resulting electronic checklist screen is shown below. In this example, a very simple trigger event needed only two team members to review the items. As each major category of the PSSR database items was marked not applicable, it collapsed to show this. For the one major category selected, only the one applicable sub-item (P&ID revision to show the local gauge's range change) remains visible and records the PSSR completion.

TABLE 6-1 Example of collapsing an electronic PSSR checklist

ANYCHEM CO. – ELECTRONIC PRE-STARTUP SAFETY REVIEW			
Date: 4/20/07	PSSR Team Leader: A. Engineer		
Facility/Process Equipment Reviewed: Replaced local pressure indicator (PI-123) displaying a 0 – 50 psig range on cooling water pump (P-456) outlet with a pressure indicator displaying a 0 – 150 psig range provided by the original equipment manufacturer using identical materials of construction. Details of the replacement are logged in the electronic work order history.			
Type of Startup	Modified Process/Restart: X		
Recommendations	None. Physical inspection performed 4/18/07		
Item No.	Description	Initials	Date Resolved
Not applicable	Not applicable	Not applicable	Not applicable

ANYCHEM CO. – ELECTRONIC PRE-STARTUP SAFETY REVIEW				
PSSR Completion Summary: The following issues have been resolved and the undersigned believe the process/facility is ready for startup.				
<ul><li>The construction and equipment meet design specifications.</li><li>Safety, operating, maintenance, and emergency procedures are in place and are adequate.</li><li>For new facilities, the initial process hazard analysis (PHA) has been performed and recommendations have been resolved.</li><li>Training of each employee involved in the operating process is complete.</li><li>Changes made to modify the process/facility have been reviewed and authorized by the facility management of change program.</li></ul>				
Confirmation by PSSR Team Members				
Name/Title			Date	
A. Engineer	Process Engineer		4/19/07	
D. Millwright	Maintenance Technician		4/19/07	
PSSR ITEM CATEGORY		Applicability		
PROCESS HAZARD ANALYSIS ISSUES		Not applicable		
DOCUMENTS IN PLACE AND APPROVED		Not applicable		
PROCESS TECHNOLOGY INFORMATION		Applicable		
Item	Responsibility	Projected Completion Date	Actual Completion Date	Initials
P&IDs	A. Engineer	4/13/07	4/10/07	A.E
MECHANICAL INTEGRITY INFORMATION		Not applicable		
INSTRUMENTATION & CONTROLS INFORMATION		Not applicable		
TRAINING REQUIREMENTS		Not applicable		
GENERAL FACILITY REQUIREMENTS		Not applicable		
CONTROL ROOM REQUIREMENTS		Not applicable		
PROCESS CONTROL SYSTEMS		Not applicable		
ENVIRONMENTAL REQUIREMENTS		Not applicable		

6. METHODOLOGIES FOR COMPILING AND USING PSSR CHECKLISTS

This approach allows the user to see only what is applicable, what items are scheduled to be reviewed, and when it is complete. Again, it may be tied to an automated action item tracking system that could be programmed to:

- Send reminders to the assigned parties when a review action is due
- Send the PSSR team leader updates on progress at a predetermined period for complex reviews
- Send the area management team reports on completeness of PSSRs within their operating area

6.3.2 Expand the Checklist for Complex PSSR

The same electronic checklist for a more complex PSSR in progress can be expanded as each applicable data field is completed. The example below shows one way this might be achieved using a customized PSSR software application.

TABLE 6-2 Example expansion of an electronic PSSR checklist in progress

ANYCHEM CO. – ELECTRONIC PRE-STARTUP SAFETY REVIEW			
Date: 4/20/07	PSSR Team Leader: A. Engineer		
Facility/Process Equipment Reviewed: Proprietary chemical addition system installation MOC ID No. 54321.			
Type of Startup	Modified Process/Restart: X		
Recommendations	Click box to expand list ☐		
Item No.	Description	Initials	Date Resolved
54321-1 through 54321-81 Click box to expand list ☐			
PSSR Completion Summary: The following issues have been resolved and the undersigned believe the process/facility is ready for startup.			
• The construction and equipment meet design specifications.			
• Safety, operating, maintenance, and emergency procedures are in place and are adequate.			
• For new facilities, the initial process hazard analysis (PHA) has been performed and recommendations have been resolved.			
• Training of each employee involved in the operating process is complete.			
• Changes made to modify the process/facility have been reviewed and authorized by the facility management of change program.			

Confirmation by PSSR Team Members		
Name/Title		Date
A. Engineer	Area process engineer	OPEN
A. Instrument	I&E Technician	OPEN
D. Millwright	Maintenance Technician	OPEN

ANYCHEM CO. – ELECTRONIC PRE-STARTUP SAFETY REVIEW					
D. Trainer	HR Department		OPEN		
A. Safety.	EHS/IH Technician		OPEN		
PSSR ITEM CATEGORY		Applicability			
PROCESS HAZARD ANALYSIS ISSUES					
Item	Responsibility	Projected Completion Date	Actual Completion Date	Initials	
Pre-startup action items closed	D.T.	3/15/07	OPEN		
DOCUMENTS IN PLACE AND APPROVED					
Item	Responsibility	Projected Completion Date	Actual Completion Date	Initials	
Operating Procedures	D.T.	3/15/07	OPEN		
Maintenance Procedures	D.T.	3/15/07	OPEN		
Safety Procedures	A.S.	3/15/07	OPEN		
Other: Process Overview	D.T.	3/15/07	OPEN		
PROCESS TECHNOLOGY INFORMATION					
Item	Responsibility	Projected Completion Date	Actual Completion Date	Initials	
P&IDs	A.E.	3/15/07	OPEN		
Process Flow Diagrams	A.E	3/15/07	OPEN		
MSDS/Chemical hazards	A.S.	3/15/07	OPEN		
Safe Operating Limits	A.E.	3/15/07	OPEN		
Process Description/Chemistry	A.E.	3/15/07	OPEN		
Equipment Lists	D.M.	3/15/07	OPEN		
Heat & Material Balance	A.E.	3/15/07	OPEN		
Relief Valve Summary List	A.E.	3/15/07	OPEN		
Instrument Safety System List	A.E.	3/15/07	OPEN		
Mechanical Safety System List	A.E.	3/15/07	OPEN		
Piping Schedule	A.E.	3/15/07	OPEN		

6. METHODOLOGIES FOR COMPILING AND USING PSSR CHECKLISTS 87

ANYCHEM CO. – ELECTRONIC PRE-STARTUP SAFETY REVIEW				
Maximum Intended Inventory	A.E./A.S.	3/15/07	OPEN	
MECHANICAL INTEGRITY INFORMATION				
Item	Responsibility	Projected Completion Date	Actual Completion Date	Initials
Rotating Equipment Records	D.M.	3/25/07	OPEN	
Electrical Equipment Records	D.M.	3/25/07	OPEN	
Inspection Records	D.M.	3/25/07	OPEN	
Relief Valve Records	D.M.	3/25/07	OPEN	
Equipment Records	D.M.	3/25/07	OPEN	
Equipment Specifications	D.M.	3/25/07	OPEN	
Electrical 1 line drawings	D.M.	3/25/07	OPEN	
INSTRUMENTATION & CONTROLS INFORMATION				
Item	Responsibility	Projected Completion Date	Actual Completion Date	Initials
DCS or PLC Documentation	A.I.	3/25/07	OPEN	
DCS Screen Revisions	A.I.	3/25/07	OPEN	
TRAINING REQUIREMENTS				
Item	Responsibility	Projected Completion Date	Actual Completion Date	Initials
Operator/Maintenance Process Overview Training	D.T.	4/13/07	OPEN	
Operator Procedure Training	D.T.	4/13/07	OPEN	
Maintenance Procedure Training	D.T.	4/13/07	OPEN	
HAZCOM Training Certified	D.T.	4/13/07	OPEN	
Fire Department Training	D.T.	4/13/07	OPEN	
GENERAL FACILITY REQUIREMENTS				
Item	Responsibility	Projected Completion Date	Actual Completion Date	Initials

ANYCHEM CO. – ELECTRONIC PRE-STARTUP SAFETY REVIEW				
Construction/Equipment Meet Design Specifications	A.E.	4/13/07	OPEN	
Ladder/Platforms Meet OSHA Standards	A.S.	4/13/07	OPEN	
Handrails/Toe boards/Walkways Meet OSHA Standards	A.S.	4/13/07	OPEN	
Electrical Systems In Service	A.I.	4/13/07	OPEN	
Lighting Adequate	A.S.	4/13/07	OPEN	
Safe Access to Block Valves, gauge glasses, and Bleeders	A.S.	4/13/07	OPEN	
Electrical meets OSHA Code	A.E.	4/13/07	OPEN	
Fire Alarm Systems Working/Tested	A.S.	4/13/07	OPEN	
Fire Extinguishers In Place	A.S.	4/13/07	OPEN	
CONTROL ROOM REQUIREMENTS				
Item	Responsibility	Projected Completion Date	Actual Completion Date	Initials
Operator Controls Operational	A.E.	4/13/07	OPEN	
Communications Operational	NA	4/13/07	OPEN	
Electrical Meets OSHA Code	A.E.	4/13/07	OPEN	
Alarm Panels In Service	A.E.	4/13/07	OPEN	
Fire Alarm System In Service	A.E.	4/13/07	OPEN	
Protective Clothing In Place	A.S.	4/13/07	OPEN	
TV Monitors Operational	A.E.	4/13/07	OPEN	
Graphic Control Panel Operational	A.E.	4/13/07	OPEN	
Computer Control Systems Operational	A.E.	4/13/07	OPEN	

6. METHODOLOGIES FOR COMPILING AND USING PSSR CHECKLISTS

ANYCHEM CO. – ELECTRONIC PRE-STARTUP SAFETY REVIEW					
Process Safety Information In Place	A.E.	4/13/07	OPEN	
PROCESS CONTROL SYSTEMS				
Item	Responsibility	Projected Completion Date	Actual Completion Date	Initials
System Problems Cleared	A.I.	4/13/07	OPEN	
Alarms & Trips at Proper Settings	A.I.	4/13/07	OPEN	
Field Switch Alarms & Trips at Proper Settings	A.I.	4/13/07	OPEN	
Graphics Correct	A.I.	4/13/07	OPEN	
Required Loops Checked/Operational	A.I.	4/13/07	OPEN	
Gauges In Place & Operational	A.I.	4/13/07	OPEN	
Instrumentation Orientated for Easy Reading	A.I.	4/13/07	OPEN	
Control Valves Tested/Calibrated	A.I.	4/13/07	OPEN	
ENVIRONMENTAL REQUIREMENTS				
Item	Responsibility	Projected Completion Date	Actual Completion Date	Initials
Verify equipment and piping components have been identified for tagging in the leak detection and repair program	A.S.	4/13/07	OPEN	
Verify process vents in hazardous air pollutant service are not vented to the atmosphere	A.E.	4/13/07	OPEN	

7
CONTINUOUS IMPROVEMENT

It could be easy to overlook this aspect of the pre-startup safety review (PSSR) program at a facility. However, PSSR performance can benefit from the same techniques used to improve product quality and productivity performance for a company. In actuality, the return on investment could even be greater from seeking continuous improvement for the PSSR element if it ensured new process aspects were implemented more safely and in a way that reduced rework and post-startup problems.

This chapter examines some ways to help ensure the PSSR system at a facility reflects the changing needs of the processes, culture, and resources as they change.

7.1 DIAGNOSING PSSR SYSTEM ISSUES

How can issues with a PSSR system be diagnosed? As with any human illness, the first step is to look for symptoms. Chapter 7 – *Continuous Improvement* provides some diagnostic tools which provide a clear-cut, systematic, and well-documented approach to finding these symptoms.

The following table summarizes some of the problems that may arise during a formal audit of the PSSR element. However, every employee should be encouraged to look for these kinds of issues each time they:

- encounter a trigger event,
- evaluate it for the risk level and PSSR approach,
- select the applicable checklist items for review, and
- complete the PSSR steps indicated for that event.

Think about each maintenance work order. Does it need a PSSR? What about operations changes? Do they need a PSSR? Remember that there is no need to wait for a scheduled audit to evaluate your system. Each use of an element of PSM is an opportunity for improvement.

TABLE 7-1 Typical PSSR issues from formal or informal audits

	PSSR ISSUES
1.	Are there written procedures for performing PSSR and are they clear and understandable?
2.	Has the PSSR team leader been trained on the site PSSR procedure?
3.	Were the necessary skills available either on the PSSR team or readily available to the PSSR team when needed?
4.	Have PSSR activities and reviews been recorded and documented for the sampled trigger events?
5.	Can the PSSR records be easily retrieved?
6.	Were the review techniques indicated by a risk-based approach applied thoroughly?
7.	Were trigger events not identified for initial PSSR evaluation?
8.	How often did a PSSR delay its associated startup?
9.	Were the identified PSSR-related startup delays beneficial to helping assure long-term safety performance?
10.	Was there a PSSR management system failure or other related management system failure?
11.	Did the PSSR hardcopy checklist or electronic database system work properly? (For example, was data ever lost or did the system hinder completion?)
12.	What other resources, techniques or tools could be used to make future similar PSSRs more effective?
13.	Were the PSSR forms completed for each trigger event?
14.	Were the scheduled PSSR compliance audits completed on time?
15.	Did scheduled PSSR compliance audits identify issues? (And if so, were they addressed?)
16.	Did scheduled PSSR audits identify good practices? (And if so, were they communicated and cultivated throughout the organization?)
17.	Is PSSR management system functioning properly?
18.	Is the PSSR tracking - follow - resolution system working?

7.2 TRAINING AND COMMUNICATION

Achieving continuous improvement implies issues (symptoms) will be found, solutions will be diagnosed, and corrective plans will be designed and implemented. The real test of a person assigned to address PSSR continuous improvement issues is to design and implement solutions that prevent a problem's recurrence. This is where training and communication come into the formula.

So much of the PSSR process relies upon consistent high-level human performance that training the affected workers in the initial system, as discussed in previous chapters, is essential. But when looking for continuous improvement, we need to evaluate re-training on occasion to keep people current and update our training on PSSR for special issues. Revising and re-implementing training for employees may be useful when a continuous improvement issue causes:

- the PSSR procedure to be revised,
- the PSSR checklist or PSSR electronic database to be revised,
- the PSSR software application user interface is changed (if one is used), or
- any legal requirements a facility is subject to are revised.

In many cases, an issue that can enhance the PSSR system performance may be so simple that full-blown training may not be required. In these cases, simply communicating the problem and how to avoid it can work. Opportunities actually arise from these communication needs. For example, when a PSSR continuous improvement issue is one that can be addressed with simple communication, it could provide an agenda discussion item for a safety meeting, a shift turnover topic for hourly workers, or an opportunity for the plant safety staff to transmit a brief e-mail reminder – also serving as a general safety consciousness raising tool. Communicating good practices found during PSSR audits can also be a way to praise groups or individuals who established the good practice and deserve recognition.

7.3 EXAMINE EXCESSES AS WELL AS DEFICIENCIES

It is important to note that any continuous improvement effort should also look at efficiency of the process it is evaluating. For example, ask questions like:

- Are we spending time on PSSR activities with a low return on our efforts?
- Are there aspects of our specific PSSR work process that can be streamlined or de-bottlenecked?

Be cautious in cutting out a practice to improve efficiency simply because it doesn't apply to a specific trigger event. That practice may be extremely important when reviewing others. But do look at possibly modifying the PSSR procedure or checklist to help users determine when efficiencies can be gained.

7.4 WHY REFINE, IMPROVE, UPGRADE, OR REDESIGN?

The concept of pre-startup safety review and operational readiness inspection is not new. Most sites have had some type of work process addressing the concept in

7. CONTINUOUS IMPROVEMENT

place for many years. So why should the PSSR process at a facility be modified? Some examples of events that may encourage a revision are presented below.

7.4.1 Workforce Reductions

Reductions in force can occur for many reasons; market changes, technology changes, and other drivers can influence the size of the workforce. Be aware of your PSSR administrative procedure and how it assigns duties. Often management and technical positions may disappear or be renamed.

A general consideration is the overall availability of personnel. If your PSSR program requires PSSR teams to be of a certain size, a workforce reduction may guide the revision of that aspect of the procedure.

7.4.2 Company Restructuring

When a company moves business units from a site, changes its management organization, or otherwise restructures operations, there is a possibility that the PSSR element will be affected.

The PSSR checklist or electronic database may contain items that no longer apply if certain raw materials, intermediates, or end products are removed from the on site. The PSSR work process itself may change if management titles change or responsibilities are reassigned. It is not uncommon to find these issues in any PSM element after such a restructuring.

7.4.3 Acquisitions, Mergers, and Divestiture

Similar to company restructuring, acquisitions, mergers, and divestitures can affect the PSSR work process. Any major change in the characteristics of the site should be evaluated for the impact on PSSR and associated elements of the process safety management system.

7.4.4 Regulatory Changes

As mentioned in the training and communication section of this chapter, regulatory changes can impact the entire PSM or RMP system at a facility. Seeking continuous improvement implies that an organization will foster awareness of the external and internal requirements it must meet in order to operate within the law. Remember, introducing a new chemical to the facility may cause the site to fall under more stringent, or in some cases, less stringent regulatory scrutiny and requirements.

7.4.5 Changes in Process Risk

Many organizations are revisiting their chemical processes to determine whether inherently safer design approaches can help them reduce or remove certain hazards and associated risks from their facilities. In other cases, new technology and advances in management science may allow a company to test or permanently adopt materials or equipment that increase the hazards and risks present, but are now manageable when added layers of protection are used. Whether process risk is decreasing or increasing, the PSSR work process and checklist/database items should be evaluated to ensure they still help PSSR teams perform effective reviews.

7.5 UPGRADING THE SYSTEM

When a continuous improvement effort at a facility results in a consensus to upgrade the system, check to see if the following characteristics of the upgrade are considered:

- **Drafting the upgrade** – It is a good practice to involve more than one person in drafting the change to a programmatic document such as a PSSR work process. Consider discussing the proposed upgrades with a team of knowledgeable subject matter experts (SMEs) during the draft phase. These can be persons involved in the continuous improvement effort or experienced PSSR team members.
- **Reviewing the upgrade** – Identify the appropriate reviewers for this document. The facility document control system or PSM/RMP administrative procedures may provide some guidance. The review process for any procedure is an excellent opportunity for employee participation. Consider having hourly, technical and management personnel involved in the review.
- **Resolving Comments** – The primary person responsible for drafting the upgrade may be able to perform this step alone if the comments received are clearly written and straightforward. However, if there are any questions as to the reviewer's intent, consider discussing their comments and possibly checking with a subject matter expert.
- **Approval Review** – Again, the facility document control system, or PSM/RMP administrative procedures may provide some guidance. Typically a document category has one management representative assigned as its owner. That might be the site manager for a programmatic document like the PSSR work process. In some facilities, it may be the safety manager. In any case, allow them to approve the document or comment, then recycle it back to the writer for resolution and resubmit it for approval.
- **Publishing** – Once the upgraded PSSR procedure is approved, publish it in accordance with the site document control work process.

Training on the newly upgraded pre-startup safety review program may need to be provided. At a minimum, the upgrades should be communicated to those persons affected by the changes.

7. CONTINUOUS IMPROVEMENT

A good practice many facilities have adopted is the use of a standard protocol for performing self- and third-party audits of their process safety management system or RMP prevention program performance. This chapter provides some examples of:

- General management system element auditing applicable to any management system
- Guidance for a company's first, second, or third party audit
- U.S. OSHA inspector protocols for conducting a PSSR audit

7.6 EXAMPLE PSSR PERFORMANCE AND EFFICIENCY METRICS

This section lists possible metrics gained from published guidelines and industry practices involving PSSR performance and efficiency.

7.6.1 PSSR Performance Indicators

PSSR performance measures that explicitly identify key indicators can be used to assess system performance on a near-real-time basis and at a more reasonable expenditure of effort. The following is a list of several indicators that may be relevant to many PSSR systems. Monitoring these key indicators can help a company or facility detect when deviations when implementing a PSSR system before these deviations can cause accidents. The sensitive indicators for a specific PSSR system will depend upon a variety of factors, including the specific PSSR system design and the availability of PSSR records and data. Some indicators can be used individually to help evaluate system performance, while other indicators must be used in combination with others. Consider measuring and analyzing the following items:

- The number of incidents that occur during startup – A high number or increasing rate might indicate that PSSRs are not being conducted in a careful manner or that MOC was not performed well
- The number of spurious shutdowns after startup – A high number or increasing rate might indicate that PSSR activities were not effective
- The number of improperly assembled pieces of equipment found during PSSRs – A high number would indicate that PSSRs were effective
- The number of personnel trained prior to startup – A high number or percentage might indicate that PSSR activities were being performed on schedule
- The duration of startup – A high number might indicate that PSSRs were not performed well
- The amount of off-spec product or loss of raw material due to startup problems – A high number might indicate that PSSR activities were not performed well or that MOC activities did not address quality hazards

- The number of people trained per year on PSSRs – A high number or percentage would indicate an active PSSR program
- The number of startups for which PSSRs were not performed – A high or increasing number would indicate that the PSSR program was being circumvented and that remedial awareness training might be in order
- The number of PSSRs for which authorizations to restart were not found – A high number or percentage would indicate that PSSRs were not being finished off or documented properly
- The number of startups deferred due to problems found during PSSRs – A high or increasing number would indicate that the MOC process was not identifying all potential problems/hazards
- The number of field revisions made during startup to deal with issues not discovered during the PSSR – A high or increasing number would indicate poor PSSR performance
- Number of action items completed after startup – A high percentage would indicate a PSSR program that was vigilant in follow-up activities

7.6.2 PSSR Efficiency Indicators

The following three items might assist in helping improve performance of an already well-run PSSR program. To improve efficiency, consider measuring the following items:

- The number of PSSRs performed per month/year – A high or increasing number would indicate an active program
- The average amount of calendar time taken from PSSR to completion of all action items – A high or increasing number might indicate the need to improve efficiency or that the MOC program was effective
- Person-hours expended on PSSRs – A high or increasing number may indicate that efficiency improvement or more resources are needed

7.7 AUDIT FREQUENCY

For those sites covered, a full compliance audit of the PSM/RMP management system must be completed every three years.

One way to facilitate the audit process is when individual elements of PSM or the RMP prevention program can be scheduled for audit by quarter to ensure all elements of the PSM system are audited every three years. This continuous auditing cycle maintains employee awareness of the overall PSM system and reduces personnel requirements.

7. CONTINUOUS IMPROVEMENT

Another approach is to do a periodic audit of 5 to 10 management of change packages to evaluate the effectiveness of the MOC/PSSR system. This provides a quality control metric for comparison over time.

7.8 QUALIFICATION CONSIDERATIONS FOR PSSR AUDITORS

Audit team members can be assigned by the person designated as responsible for auditing the element. PSSR element audit teams can include members of the facility management, as well as staff and hourly employees. Other facilities within the organization, especially a corporate PSM/RMP group may provide audit team members to participate in or lead the audits as well. This practice provides a benefit of sharing both good practices and typical deficiencies throughout the entire organization. Other company facilities may also provide audit team members to participate in or lead the audits. Some companies use consultants.

The facility leadership team may designate persons knowledgeable about the area or process to participate in the PSSR audit based upon their current job assignment or past experience.

Although others may be designated to assist in the audit, a reasonable dissemination of the responsibilities to complete an audit for each element of PSM follows. Some companies find that these responsibility assignments logically divide the work among the positions or departments most affected by or responsible for a given element. Consider establishing a document retention guideline for PSSR forms. Some companies use 5 years to match their 5 year accident history requirement when they are covered by the EPA risk management program rule.

All audit team members should have the following attributes:
- Familiarity with the facility's PSM/RMP compliance plan or organizationally driven internal PSSR requirements
- A basic knowledge of the PSM/RMP elements, especially the pre-startup safety review element
- Attendance at a process safety management training course, either in-house or a generic course offered by a reputable educational organization such as AIChE
- Training on audit techniques
- At least one member of the PSSR audit team should be knowledgeable in the processes in the specific facility being audited.

Audits can be internal or external:

- Internal typically means the facility audits itself with no outside influence. This is termed a first-party audit.
- An internal second-party audit typically means corporate groups or other plants are involved. Typically a second-party audit involves a corporate representative or appropriate positions from another site within the company. This method brings outside points of view and experience to a site and can be very helpful in sharing best practices as well as communicating issues to other corporate facilities.
- External third-party audits include an outside party to examine the system. This can be a contractor, a regulatory agency, or a customer. A third-party audit typically means an outside consultant or organization comes in to look at your system and compare it with standards or regulations from yet another standpoint.

7.9 SAMPLE PSSR AUDIT PROTOCOLS

The following are examples of pre-startup safety review protocols. These help ensure consistency in the audit process over the life of a facility's operation.

One approach is a generic audit protocol for any management system element. It could apply to an environmental management system, a process safety management system, a quality management system, or an integrated business management system element. This approach looks at three primary areas:
- The element administrative procedures or policy's existence and adequacy
- A field audit of whether or not the procedure is being used
- An audit of any supporting documentation which may (or may not) be required by the element

7. CONTINUOUS IMPROVEMENT

Another approach is to customize a protocol for a specific facility's PSSR program. The following example shows wording that might appear in a PSSR audit protocol guidance document. Table 7-2 – *Example guideline wording for a PSSR element audit protocol* shows how a facility might implement this type of protocol

If a facility is covered under the OSHA PSM regulation, it may determine that internal audits using the current guidelines for OSHA auditors is one way to evaluate their PSSR system. At the time of this writing, U.S. OSHA provides the guidance presented in Table 7-3 – *Excerpted PSSR language from OSHA CPL 2-2.45A CH-1* to its inspectors in OSHA Instruction CPL 2-2.45A CH-1 *Subject: 29 CFR 1910.119, Process Safety Management of Highly Hazardous Chemicals -- Compliance Guidelines and Enforcement Procedures.*

The most recent version of this document establishes uniform policies, procedures, standard clarifications, and compliance guidance for enforcement of the standard for Process Safety Management of Highly Hazardous Chemicals, 29 CFR 1910.119 ("PSM standard"), and amendments to the standard for Explosives and Blasting Agents, 29 CFR 1910.109. Of course, facilities using this as a guide should regularly check for updates to the instruction. Note that interviews with employees are an essential part of the audit protocol.

TABLE 7-2 Example guideline wording for a PSSR element audit protocol

Physical Pre-startup Safety Review – Audit Protocol
Physical inspections shall be guided by prewritten checklists and involve extensive field inspections. The attributes of a "good practice" checklist are as follows:
• A clear front sheet stating what equipment has been inspected, when, and by whom
• A statement that the installation is consistent with design specifications
• Written statements that the PSSR team concludes that the facility is safe to start up after certain recommendations have been satisfactorily resolved.
• A record of recommendations (if any) with timing and responsibility for completion either before or after start-up.
• A check that the following basic safety and occupational health areas have been appropriately addressed:
o General safety
o Machinery safety
o Ergonomics
o Occupational health
o Operating procedures and safe work practices
o Training and performance
o Contractor safety
o Interlocks (both safety & non-safety) and alarms
o Heat transfer media
o Highly toxic materials
o A check that other relevant safety, health and environmental topics have been addressed
o Environmental
o Community awareness and emergency response
o Electrical safety
o Fire protection

7. CONTINUOUS IMPROVEMENT

TABLE 7-3 Excerpted PSSR language from OSHA CPL 2-2.45A CH-1

CPL 2-2.45A CH-1 Subject: 29 CFR 1910.119, Process Safety Management of Highly Hazardous Chemicals – Compliance Guidelines and Enforcement Procedures
1910.119(i): PRE-STARTUP SAFETY REVIEW
I. PROGRAM SUMMARY
The intent of this paragraph is to make sure that, for new facilities and for modified facilities when the modification necessitates a change to process safety information, certain important considerations are addressed before any highly hazardous chemicals are introduced into the process. Minimum requirements include that the pre-startup safety review confirm the following: <ul><li>construction and equipment is in accordance with design specifications;</li><li>safety, operating, maintenance, and emergency procedures are in place and adequate;</li><li>for new facilities, a PHA has been performed and recommendations resolved or implemented;</li><li>modified facilities meet the requirements of paragraph (1), management of change; and training of each employee involved in the process has been completed.</li></ul>
II. QUALITY CRITERIA REFERENCES
A. 1910.119(i): Pre-startup Safety Review
B. 1910.119(l): Management of Change
III. VERIFICATION OF PROGRAM ELEMENTS
A. Records Review
1. Has a pre-startup safety review been performed for all new facilities and for modified facilities when the modification is significant enough to require a change in process safety information? [Criteria Reference .119(i)(1)]
2. Do pre-startup safety reviews confirm that prior to the introduction of highly hazardous chemicals to a process: [Criteria Reference .119(i)(2)]<ul><li>Construction and equipment is in accordance with design specifications?</li></ul>
<ul><li>Safety, operating, maintenance, and emergency procedures are in place and adequate?</li></ul>
<ul><li>For new facilities, a PHA has been performed and recommendations resolved or implemented before startup?</li></ul>
<ul><li>Modified facilities meet requirements of paragraph (l)?</li></ul>
<ul><li>Training of each employee involved in operating the process has been completed?</li></ul>

CPL 2-2.45A CH-1 Subject: 29 CFR 1910.119, Process Safety Management of Highly Hazardous Chemicals – Compliance Guidelines and Enforcement Procedures
B. On-site Conditions
1. Do observations of new or modified facilities indicate that prior to the introduction of highly hazardous chemicals:
Construction and equipment is in accordance with design specifications? Safety, operating, maintenance, and emergency procedures are in place and adequate?
C. Interviews
1. Based on interviews with a representative sample of operators, maintenance employees, and engineers, can it be confirmed that the construction and equipment are in accordance with design specifications prior to introducing highly hazardous chemicals to a process? [Criteria Reference .119(i)2(i)]
2. Based on interviews with a representative sample of operators, maintenance employees, and engineers, are safety, operating, maintenance, and emergency procedures in place prior to introduction of highly hazardous chemicals into a process? Are these procedures adequate? [Criteria Reference .119(i)2(ii)]
3. Based on interviews with a representative sample of operators, maintenance employees, and engineers, is a PHA performed and are recommendations resolved prior to a startup that introduces highly hazardous chemicals into a new process? [Criteria Reference .119(i)2(iii)]
4. Based on interviews with a representative sample of operators, maintenance employees, and engineers, do modified facilities meet requirements of paragraph (1), Management of Change prior to introducing a highly hazardous chemical? [Criteria Reference .119(i)2(iii)]
5. Based on interviews with a representative sample of operators, is training completed for each employee involved in operating the process prior to the introduction of a highly hazardous chemical? [Criteria Reference .119(i)2(iv)]

7.10 ADDRESSING AUDIT RESULTS

Each of the methods presented implies some form of reporting of the audit results. Many organizations like to use the audit process to identify discrepancies in compliance with outside or internal requirements, but also as a way to identify and spread good practices or innovative approaches to implementing the element being reviewed.

In the case of PSSR, audit action items should be tracked with the appropriate PSM/RMP action-item tracking methods used for other elements at a site. They should be assigned to a specific person with an estimated due date for completion. The best systems also allow notes for capturing history of the action item.

Management can use the open, late, and outstanding action item numbers as a metric for PSM/RMP activity and compliance levels. It is a very useful way to measure the company's implementation and appreciation of the philosophy behind process safety.

7.11 SUMMARY

Upgrading the PSSR work process procedure brings the continuous improvement process full circle. As the newly revised program is implemented, the users can identify any continuous improvement related change and submit it for consideration.

7.12 REFERENCES

7-1 OSHA Instruction CPL 2-2.45A CH-1 September 13, 1994 Directorate of Compliance Programs, *Subject: 29 CFR 1910.119, Process Safety Management of Highly Hazardous Chemicals – Compliance Guidelines and Enforcement Procedures.*

APPENDIX A – PSSR CHECKLIST EXAMPLES

Effective pre-startup safety review programs can use many different approaches to their checklist design. Several examples follow. Compare these against each other and evaluate their similarities and differences.

These examples can also serve as a reference for the various types of checklist items a facility might use as example to populate their own PSSR checklists, PSSR templates, or electronic PSSR software database items.

PSSR Checklist Example A-1

Description of System/Area Under Review	Date	Time

List of Participants	Circulation: those present plus...

Comments

Item No.	Recommendation (Type Action Below or 'not applicable')	Department / Responsible Person	Completed Date
1.	Review HAZOP list for applicable actions		

Item No.	Recommendation (Type Action Below or 'not applicable')	Department / Responsible Person	Completed Date
2.	Review CHAZOP list for applicable actions		
3.	Review punch list for applicable actions.		
4.	Noise level monitoring (document)		
5.	Exposure monitoring (document)		
6.	Emissions permits		
7.	Other operating permits		
	System Checkout Items		
8.	Complete/document quality assurance program:		
	a) Received equipment=purchased		
	b) Adherence to design drawings/specifications		
	c) Materials of construction		
	d) Workmanship		
	Equipment components checkout/commissioning (document)		
9.	Loop and interlock/permissive/alarm: set point adjustment and testing (document results)		
10.	Instrument calibrations done and documented		
11.	Validation protocol developed, approved and executed		
12.	Pressure testing done and documented		
13.	System cleaned and flushed		
14.	Fire protections systems inspected, acceptance-tested documentation provided		
15.	Ventilation systems balanced		
16.	Performance checkout of local exhaust systems		
17.	Vibration measurements/documentation		
18.	Field radiography done and documented as required		

APPENDIX A

Item No.	Recommendation (Type Action Below or 'not applicable')	Department / Responsible Person	Completed Date
19.	Baseline data, such as thickness readings for pressure vessels/piping		
20.	Other baseline data		
21.	NDT reports		
22.	Water batching (document)		
	Engineering Design Documentation		
23.	Narrative of control philosophy/Sequence of operations documentation		
24.	Instrument index		
25.	Instrument loop diagrams		
26.	Interlock (safety & non-safety) descriptions		
27.	As-built drawings: P&IDs, electrical, piping, mechanical		
28.	Tabulation of process alarms, interlocks/permissive descriptions and trips with settings (P&ID)		
29.	Review and approval of fire protections systems design		
30.	Piping specifications (P&I/mechanical drawings)		
31.	Pressure relief device sizing calculations		
32.	ASME code pressure vessels U-1 data sheets (to Maintenance)		
33.	Reference codes/standards for facility design		
34.	Reference codes/standards for facility installation		
35.	Design codes for specific equipment		
36.	Welder certification		
37.	NDT certification		
38.	Electrical classification drawings		
39.	Electrician certification for classified areas		
	Maintenance Items		

Item No.	Recommendation (Type Action Below or 'not applicable')	Department / Responsible Person	Completed Date
40.	Spare parts lists developed *List equipment:*		
41.	Stock required spare parts		
42.	Equipment manuals/specifications to Maintenance and Operating Department		
43.	Vendor prints		
44.	Submit preventive maintenance requests (PM's)		
45.	Service contracts in place		
46.	Train maintenance personnel and document		
	Operational Items		
47.	Develop SOP and special procedures – incorporate safety and operational issues:		
48.	Complete change control authorizations		
49.	Develop/issue operating procedures:		
	a) Initial start-up		
	b) Normal start-up		
	c) Normal operation		
	d) Normal shutdown		
	e) Emergency operations including emergency shutdown		
	f) Start-up following emergency shutdown		
	g) Start-up following turnaround		
	h) Non-routine procedures (equipment clean-out, equipment preparation for maintenance)		
	i) Auxiliary equipment operation		
50.	Train operating personnel and document		
51.	Provide technical coverage		

APPENDIX A

Item No.	Recommendation (Type Action Below or 'not applicable')	Department / Responsible Person	Completed Date
52.	What is the probability of containment failure? What are the subsequent consequences? Review spill containment, rainwater runoff and fire water containment.		
	Electrical Issues		
53.	Panel access, clearance around panel, keys, etc.		
54.	Cable entry section – glanding, housekeeping and other issues		
55.	Panel power source identified? Multiple feeds?		
56.	Space heaters required?		
57.	Lighting in panels		
58.	Canopies etc. for outdoor panels		
59.	Protection settings available?		
60.	Panel board schedules		
61.	Spare capacity available?		
62.	PPE available? HV mats, gloves		
63.	Emergency/Standby power required?		
64.	VSD settings available – factory/commissioning		
65.	Room access, permitting, maintainability, ingress/egress of equipment		
66.	Pockets for drawings in panels?		
67.	Voltage warning labels?		
68.	Ventilation sufficient?		
69.	Suitability for area - IP rating, GMP suitability, hazardous area classification.		
	Relief Devices		
70.	Are there standard markings on the P&ID?		
71.	Do relief lines vent to a safe location?		
72.	Are relief lines and relief devices secured and adequately sized?		

Item No.	Recommendation (Type Action Below or 'not applicable')	Department / Responsible Person	Completed Date
73.	Are there any isolation valves which will inhibit operation of relief valves if closed? If YES, nominate Operations to monitor. If NO, record no.		
74.	Is there a standard operating procedure for relief devices in place?		
	For Field Verification		
75.	Is lighting sufficient?		
76.	Is emergency lighting sufficient?		
77.	Are steam pipes, valves or traps situated within proximity of people insulated?		
78.	Is all instrumentation identified and tagged?		
79.	Is there any rusted or damaged equipment?		
80.	Are swing gates installed on top of ladders or on access platforms?		
81.	Review edge protection on platforms and heights. Will it protect personnel and equipment? Is access adequate?		
82.	Do safety showers provide a hazard to people (slips), product (contamination of systems) or ingress to electrical switchgear or equipment?		
83.	Are safety showers supplied from tempered water?		
84.	Is chemical dosing within a safe proximity of people and product?		
85.	Label all pipe lines		
86.	Label all electric switches, disconnects, MCCs, control panels, cables, etc.		
87.	Label vessels (material, hazard warnings) List:		
88.	Seal wall penetrations		
89.	Pour conduit seals		
90.	Install fire extinguishers		

APPENDIX A

Item No.	Recommendation (Type Action Below or 'not applicable')	Department / Responsible Person	Completed Date
91.	Evacuation routes posted		
92.	Install appropriate area signs		
93.	Is all scaffolding and equipment removed; is general housekeeping acceptable?		
94.	If there is potential for entrapment or exposure, has an E-Stop been provided?		
95.	If full guarding is in place, has a lockable isolation device been provided?		
96.	Pest control required in room/building?		
97.	MSDS sheets required?		
98.	Sprinklers in rooms?		
99.	Single point of failure condition? Has it been identified? – Record.		

PSSR Checklist Example A-2

Pre-startup Safety Review Checklist	
Inspection Date:	
Department/Area:	
Project Number:	
Title/Equipment:	

Signatures below indicate acceptance that the equipment or project is safe and satisfactory to start-up with the exceptions noted.

Engineering / Maintenance	Date
EHS Group	Date
QA Group	Date
Manufacturing / Operations	Date
Project Engineer	Date
Process Engineering	Date

APPENDIX A

Checklist Item No.	Details (reference category/item no.)	Responsibility	Complete Sign & Date
Category A Action Items - *Items to be completed BEFORE authorization and start-up*			
1.			
2.			
3.			
4.			
5.			
Category B Action Items - *Items to be completed AFTER start-up*			
1.			
2.			
3.			
4.			
5.			
6.			
7.			
8.			
9.			
10.			
11.			
12.			
Sign below only when all punch list "before start-up" items are completed			
Authorized:	Facility/Equipment Owner Signature:		Date

PSSR ITEM NO.	CATEGORY/ITEM TO ASSESS	Not Applicable
1.1	GENERAL SAFETY	
1.1.1	Have *ALL* appropriate personnel (Operations, Maintenance, Technical, and Supervision) received adequate and appropriate training on the equipment and operating procedures?	
1.1.2	Has adequate and appropriate *PPE* (Personal Protective Equipment) been specified in the Work Procedures and/or Standard Operating Procedures. Has the PPE been provided? Have the PPE users been trained in the use of the PPE? Is the training documented?	
1.1.3	Have measures been taken to adequately guard all dangerous parts of this equipment?	
1.1.4	Has sufficient provision been made for the electrical and/or mechanical isolation of the equipment?	
1.1.5	Are points of isolation clearly marked/labeled and readily accessible?	
1.1.6	Have bump/trip hazards been properly identified and adequately marked? Have all sharp edges been removed?	
1.1.7	Has proper guarding, handrails/barriers, been provided to prevent falls?	
1.1.8	Have all hot/cold surfaces been adequately guarded to prevent burns? Are all cold surfaces adequately insulated to prevent condensation drips (slip hazards)?	
1.1.9	Are Safety Showers and Eye Wash facilities provided and adequately marked? Are the Safety Showers and Eye Wash facilities routinely inspected? Do the Safety Showers and Eye Wash facilities locations comply with Corporate guidelines? Are the Safety Showers and Eye Wash facilities readily visible and accessible?	
1.1.10	Has sufficient lighting been provided so that operation, servicing, maintenance and repair of the facilities can be carried out safely?	
1.1.11	Are notices, dials, screens, etc. for providing operational instructions, safety warnings and emergency information provided, if required, and positioned so that they are clearly visible and easily read?	

APPENDIX A

PSSR ITEM NO.	CATEGORY/ITEM TO ASSESS	Not Applicable
1.1.12	Have all overhead fixtures, for example, pipe-hangers, pipe sleeves, pipe sleeve covers, valve handles, floor opening covers, etc., which could fall or be dislodged, been properly secured?	
1.1.13	Are all of the applicable Work Permit Procedures (Confined Space Entry, Lock Out/Tag Out, Hot Work, High Work, etc.) in place? Have the Operating, Maintenance and Supervisory personnel been properly trained on the Work Permit Procedures?	
1.1.14	Has the fire protection systems been inspected by the insurance company? Has acceptance testing been completed and documented? Is there an agreed on test and inspection program for the fire protection systems (including alarm systems)?	
1.2	MACHINERY/EQUIPMENT SAFETY	
1.2.1	Has the machinery/equipment been installed so that it will be stable and secure during operation?	
1.2.2	Has all access to dangerous moving parts, or danger zones created by the equipment, been prevented by the provision of the correct guards, interlocks (both safety & non-safety) and/or barriers?	
1.2.3	Have the correct safety measures been taken to prevent any risk from hot/cold surfaces, ejection of material, failure of parts and their ejection, overheating/fire?	
1.2.4	Has safe access been provided to the equipment that requires operator and calibration and maintenance personnel access for normal operations, adjustments, service, calibration, maintenance or repair? Have slip, trip, trap, crush, entanglement, fall, bump and cut hazards been minimized?	
1.2.5	Is the equipment provided with the properly identified *START/STOP* and *EMERGENCY* controls that are positioned for safe operation without hesitation, or loss of time, and without ambiguity?	
1.2.6	Is the equipment provided with a clearly identified means to securely isolate it from *ALL* energy sources?	
1.3	ERGONOMICS	
1.3.1	Have the workstations, workplace, or equipment been constructed so that need for stooping, bending stretching, over-reaching and working over-head during operation has been eliminated or minimized?	
1.3.2	Has the need to lift, carry, push or pull heavy loads, or parts, been eliminated to the extent possible?	
1.3.3	Are all display screens, dials and *START/STOP/ EMERGENCY* buttons positioned so that they are readily visible and accessible by the operating personnel?	

PSSR ITEM NO.	CATEGORY/ITEM TO ASSESS	Not Applicable
1.3.4	Have Visual Display Screens been positioned so that interference from glare is reduced to the minimum?	
1.3.5	Have workstations been designed and equipped so that the operator can adopt a comfortable position? *(That is, able to stand, or change position and sit upright, angle at elbows and knees 90°, feet on floor.)*	
1.3.6	Does the operation of this equipment increase the risk of Upper Limb Disorder; for example, repetitive tasks, handling operations, machine paced work and prolonged operation?	
1.4	OCCUPATIONAL HEALTH	
1.4.1	Have all health risks arising from the gases, liquids, dusts, mists, biological hazards or vapors used by, contained in or emitted by this equipment been assessed? Have the health risks been eliminated or are adequate engineering controls utilized to minimize the risks?	
1.4.2	Has adequate *RPE* (Respiratory Protective Equipment) been specified in the Operating Procedures?	
1.4.3	Has the need for an Occupational Health Monitoring Programme been assessed? Has a Monitoring Programme been scheduled?	
1.4.4	Have the Operating Procedures been reviewed to take into account any additional "health hazards" which may arise from operation or maintenance of this equipment?	
1.4.5	Has adequate *LEV* (Local Exhaust Ventilation) been installed, tested, balanced and entered on an Inspection Schedule?	
1.4.6	Have adequate inspection/cleaning ports been provided on ductwork?	
1.4.7	Are relief facilities directed to a safe place away from the workplace?	
1.4.8	Has a Noise Survey been considered and a Noise Compliance Plan prepared, if required?	
1.4.9	Has all insulation been identified?	
1.4.10	Has all pipe work, tanks, and equipment containing hazardous materials been adequately labeled?	
2.0	PROCESS SAFETY	
2.1	PROCESS TECHNOLOGY	
2.1.1	Are up-to-date Material Safety Data Sheets available?	
2.1.2	Have the hazardous effects of inadvertent mixing of different materials been considered (that is, has a chemical interaction matrix been prepared/updated)?	

APPENDIX A 117

PSSR ITEM NO.	CATEGORY/ITEM TO ASSESS	Not Applicable
2.1.3	Has the process design basis been documented or updated? Has the control philosophy and sequence of operations been documented?	
2.1.4	Has the equipment design basis (for example, BPF's/P&IDs) been documented/updated?	
2.1.5	Have the recommendations from safety reviews, Process Hazards Analysis (PHA), Hazards and Operability Reviews (HAZOP), CHAZOP, or others, been implemented? Record any incomplete items.	
2.1.6	Are all relief devices shown on the P&IDs? Are standard markings used on the relief devices? Are the relief/rupture pressures included on the P&IDs?	
2.1.7	Have the pressure relief device calculations been provided? Was DIERS technology utilized to size the pressure relief devices for all pressure vessels? Does the sizing of pressure relief devices agreed with the calculated sizes? Do the calculations take into the downstream piping?	
2.1.8	Do the relief devices vent to safe locations? Is containment provided for liquids and solids released from pressure relief devices?	
2.1.9	Are there isolation valves that, if closed, will inhibit the operation of pressure relief devices? If yes, Operations must establish control plans to insure that the isolation valves cannot inhibit the operation of the pressure relief devices.	
2.1.10	Are all pressure relief devices included in the Preventive Maintenance Program? Are the inspection and testing of relief devices in accordance with local regulations?	
2.2	MANAGEMENT OF CHANGE – TECHNOLOGY/ MANAGEMENT OF SUBTLE CHANGE	
2.2.1	Has a management of change – technology/subtle change document (for example, Change of Design - *COD*) been approved?	
2.2.2	Has a test authorization been approved?	
2.2.3	Are all action items, arising from the COD, that were deemed necessary for start-up, complete?	
2.2.4	Have all changes made during construction been recorded and authorized? Have hazards evaluations (PHAs, HAZOP or CHAZOP) been done on all the changes made during construction?	
2.3	PROCESS HAZARDS ANALYSIS	

PSSR ITEM NO.	CATEGORY/ITEM TO ASSESS	Not Applicable
2.3.1	Have project PHAs been approved and a final project safety report been prepared?	
2.3.2	Are all action items, deemed necessary by the PHA team for start-up, complete?	
2.3.3	Has the project been approved as "Safe to proceed with" by the PHA team?	
2.4	QUALITY ASSURANCE	
2.4.1	Have checks and inspections been made to ensure that critical equipment is installed properly and is consistent with design specifications and vendor's recommendations (for example, alarm and interlock (safety & non-safety) tests; equipment alignment and service to process inter-connections)?	
2.4.2	Have quality assurance inspection reports, covering fabrication, assembly and installation, been completed in accordance with the project's quality assurance plan and reports filed with the equipment and design basis documentation?	
2.4.3	List specific items field checked as part of this Pre-Start-up Safety Review to ensure that:	
	The construction meets the design specifications.	
	The construction matches the drawings.	
2.4.4	Have the following documented been provided and approved: Instrument indexes and instrument loop diagrams?	
	A tabulation, including settings, of interlocks (both safety & non-safety) and trips (hardware and software), process alarms and permissive descriptions?	
	As-built drawings covering P&IDs, electrical, piping and mechanical?	
	Data sheets for pressure equipment built to ASME or equivalent codes?	
	Welder certification?	
	Non-destructive test (NDT) certifications?	
	Electrical certification for classified areas?	
2.5	MECHANICAL INTEGRITY	
2.5.1	Have maintenance procedures been approved?	
2.5.2	Have maintenance personnel been trained?	
2.5.3	Have spare parts listed been developed and entered into the parts ordering software program?	
	Are there adequate inventories of spare parts, operating supplies and maintenance materials?	
2.5.3	Have quality control procedures been approved for maintenance materials and spare parts?	

APPENDIX A

PSSR ITEM NO.	CATEGORY/ITEM TO ASSESS	Not Applicable
2.5.4	Have inspections and tests, including regulatory requirements) for the following equipment been included in a maintenance schedule?	
	Pressure vessels and storage tanks?	
	Pressure relief systems, vent systems and devices?	
	Critical controls, interlocks (both safety & non-safety), alarms and instruments?	
	Emergency devices (including shutdown systems and isolation systems)?	
	Fire protection equipment?	
	Piping systems (incl. Components, for example, valves, excess flow valves, expansion bellows) in critical service?	
	Key process-to-service tie-ins?	
	Electrical earthening, grounding, bonding?	
	MCC starters?	
	Emergency alarm and communication system?	
	Monitoring devices and sensors?	
	Pumps?	
	Lifting equipment?	
2.5.5	Has a reliability engineering analysis been considered/ completed for PSM critical equipment?	
2.5.6	Is the equipment inspected by any outside body and certificates on file (for example, CE marking, lifting equipment test certificates, pressure systems regulations, and other items)?	
2.5.7	Have all commissioning tests or inspections been identified (for example, pressure or leak tests, passivating procedures)?	
2.6	OPERATING PROCEDURES AND SAFE WORK PRACTICES	
2.6.1	Have standard operating procedures been prepared/updated and approved? Do the operating procedures cover:	
	Initial start-up? Normal start-up? Normal operations? Normal shutdowns? Emergency operations including emergency shutdowns? Start-up after emergency shutdowns? Start-up following turnarounds/prolonged shutdowns?	

PSSR ITEM NO.	CATEGORY/ITEM TO ASSESS	Not Applicable
	Non-routine procedure such as equipment clean-outs and preparation of equipment for maintenance?	
	Auxiliary equipment operations including as examples; LEV and Ventilation Systems, Heat/Cool Skids, Water (Soft, RO, WFI, Tower, etc.) Systems, Instrument and Process Air Systems, Waste Treatment Systems, Cooling (Glycol Refrigeration) Systems, Steam Generation, and others?	
	Safety and operational issues?	
	Change control procedures?	
2.7	TRAINING AND PERFORMANCE	
2.7.1	Has specific process (or job task) training been given to personnel?	
2.7.2.	Have training records been updated?	
2.8	CONTRACTOR SAFETY	
2.8.1.	Have all contract personnel been adequately trained in appropriate: chemical awareness, maintenance and operating activities and evacuation procedures?	
2.9	INTERLOCKS AND ALARMS	
2.9.1	Has the alarm/interlock (safety & non-safety) been classified and designed by the Project Team? Did the Project Team include members of the PHA team?	
2.9.2	Did the loop testing confirm that the alarm/interlock (safety & non-safety) action proved, under all conceivable failure conditions, to be fail-safe?	
2.9.3.	Prior to this PSSR, has an interlock/critical alarm Standard Operating Procedure for testing, through to the final element, been prepared and reviewed/authorized by a competent person for each new or upgraded control system?	
2.9.4.	For alarms/interlocks (both safety & non-safety) with more than one software or hardware circuit, have all possible interlock routes been tested?	
2.9.5	Has all appropriate process technology been updated (for example, interlock lists, P&IDs, logic drawings, etc.)?	
2.9.6	Does your Control System documentation adequately specify:	
	All major components and their model and serial numbers?	
	All communication cables layout and configuration?	
	Any configurable or custom settings and set-up?	
2.9.7	Has consideration been given to suitable fire detection and prevention systems for the equipment?	
2.9.8	Do you have an appropriate procedure to ensure that your software is protected (for example, routinely archived, key/password protected, etc.)?	

APPENDIX A

PSSR ITEM NO.	CATEGORY/ITEM TO ASSESS	Not Applicable
2.9.9	Has the software been properly documented and filed (for example, logic drawings, schematics, sequence/batch descriptions)?	
2.9.10	Has all software been properly validated and tested?	
2.9.11	Is there verification that the equipment does not re-start, either on the re-setting of a protective device such as an interlock (safety & non-safety), or the re-establishment of power after an outage?	
3.0	ENVIRONMENTAL	
3.1	Are all secondary containment/bonding facilities adequate?	
3.2	Are all material storage facilities adequate and appropriately labeled?	
3.3	Have adequate arrangements been made, prior to start-up, for the identification, classification and safe disposal of all waste materials?	
3.4	Have all materials, used in the system, been entered on the Area Chemicals Inventory List (or equivalent)?	
3.5	Are updated Area Spill Procedures available?	
3.6	Are material Unloading Facilities adequate and constructed in accordance with Corporate Safety, Health and Environmental Standards? Is there adequate containment (110% of truck volume) in the unloading areas for bulk liquid chemicals?	
3.7	Have the Corporate Environmental Guidelines been followed during the design stage of this project?	
3.8	Have all waste streams been identified, quantified, analyzed and minimized?	
3.9	Are all of the applicable Construction, Environmental and Operating Permits up to-date and approved?	
4.0	CAER	
4.1	Have all necessary precautions been taken to ensure that the equipment is not a source of ignition to any flammable materials, irrespective of their source?	
4.2	Are fire protection facilities adequate for example, fire extinguishers, fire walls, sprinkler systems, Alarm Boxes, etc.)?	
4.3	Are Emergency Escape Routes, including ladders, adequate and properly signposted?	
4.4	Is emergency lighting adequate?	
4.5	Is sufficient Respiratory Protective Equipment, such as Escape Sets or Self-Contained Breathing Apparatus (SCBA) available?	
4.6	Have Emergency Procedures been prepared and relevant personnel trained?	

PSSR ITEM NO.	CATEGORY/ITEM TO ASSESS	Not Applicable
4.7	Is the Community Panel advised of proposed new major projects?	
5.1	Has an Electrical Safety Checklist (Acceptance of Electrical Installations) been completed by a competent personnel?	
5.2	Has the equipment been properly installed and constructed to Corporate guidelines and local legislation, and does it meet any special installation requirements noted on the manufacturer's certificate?	
5.3	Has equipment been designed and purchased for the conditions under which it will operate (for example, hazardous areas)?	
5.4	Are all live parts adequately enclosed to prevent access?	
5.5	Does grounding and bonding comply with corporate and local standards/legislation?	
5.6	Have fuses or circuit breakers been provided which will automatically disconnect the supply?	
5.7	Are First Aid Stations, single line drawings and PPE requirements available in Motor Control Centers (MCC), Electrical Control Rooms (ECR)/Sub-stations, as appropriate?	
5.8	Have all relevant documentation and drawings (for example, P&IDs, SLDs, Schematics, equipment arrangement, I/O, logic, electrical classification and Panel Schedule drawings) been updated to reflect the current installation?	
5.9	Have all new Sub-station Breakers, MCC isolators, starters or other appropriate equipment been registered on to the Site Inspection Schedule?	
5.10	Have any electrical circuits, made redundant by this installation, been properly D&R'd?	
6.0	FIELD VERIFICATION	
6.1	Is the normal lighting adequate for normal and maintenance operations?	
6.2	Is emergency lighting sufficient?	
6.3	Are all hot and cold surfaces, which may cause burns, in the proximity of personnel insulated?	
6.4	Are all instruments, equipment and piping adequately labeled?	
6.5	Is there any rusted and/or damaged equipment?	
6.6	Are swing gates or chains installed at the top of ladders and/or on access platforms?	
6.7	Are there any gaps between platforms and equipment that could create a foot hazard?	
6.8	Is equipment and platform access adequate?	

APPENDIX A

PSSR ITEM NO.	CATEGORY/ITEM TO ASSESS	Not Applicable
6.9	Do safety showers/eye wash stations create a hazard to personnel (slips), potential for contamination of product (entry to equipment) or ingress to electrical equipment?	
6.10	Are safety showers and eye wash stations adequately marked and readily visible? Is the access to the safety showers and eye wash stations uninhibited?	
6.11	Are all pipelines labeled?	
6.12	Are all electrical switches, disconnects, MCCs, control panels, cables, etc labeled?	
6.13	Is all the equipment clearly labeled? Where required are the materials and hazards included on the labeling?	
6.14	Are wall penetrations adequately sealed?	
6.15	Are electrical conduits sealed in accordance with code requirements?	
6.16	Are evacuation routes clearly marked?	
6.17	Are fire extinguishers installed properly?	
6.18	Has the required signage been posted?	
6.19	Are emergency stops provided where there is a potential for entrapment or exposure?	
6.20	Has all scaffolding and construction equipment been removed? Is housekeeping acceptable?	
6.21	Is all required equipment guarding installed	
6.22	Does all the applicable equipment have the required CE marking displayed? Does all the applicable equipment have the required UL listing/labeling?	
6.23	Have noise-monitoring evaluations been completed? Have signs been posted where noise levels excess 85dB? Are ear-plugs available near areas exceeding 85 dB?	

PSSR Checklist Example A-3

ATTACHMENT A - PROCESS PRE-STARTUP SAFETY REVIEW CHECKLIST
AREA OR PLANT UNDER REVIEW:
DATE:
LIST OF PSSR TEAM MEMBERS:
Instructions for using this form: 1. Review the entire checklist and mark a check in column A to indicate an item or area to be included in the review. 2. If there are issues to be resolved after the initial review, complete Attachment B - *PSSR Potential Issue – Finding Form* 3. For each item or area with a check in column A, place a check in column B when the item or area has been satisfactorily reviewed or a potential problem has been resolved.

Column A Include	Column B Completed	Category/PSSR Item to Evaluate	
		Location and layout	
		Site Conditions	
			Drainage
			Flood control/protection
			Prevailing wind
			Air or water pollution exposures
			Other site conditions requiring attention
			Soil protection in storage, materials handling & process areas
		Nearby operations	
			Hazards from
			Hazards to
		Traffic	
			Vehicular/railroad/pedestrian
			Clearances, hazards
			Adequacy of traffic signs
		Security	
			Special requirements imposed by new facility
		Storage and handling of chemicals	

APPENDIX A 125

Column A Include	Column B Completed	Category/PSSR Item to Evaluate
		Buried pipes, tanks or chemical sewer
		Leak detection and containment
		Above ground storage tanks
		Adequate secondary containment provided
		Operating and maintenance access adequate and safe
		Adequate and accessible manways
		Unobstructed pressure/vacuum relief vents
		Manifolding of vents reviewed
		Documented vent sizing basis (process safety manuals)
		Winterization (including instrument connections)
		Adequate lighting
		Labeling, placarding of hazards
		Other installation details
		Flammable and combustible liquids
		Tank placement and spacing adequate
		Steel supports requiring fireproofing
		Flammable liquid breather vents provided with flame arrestors or conservation vents
		No flame arrestors on emergency relief vents
		Safe vent discharge locations
		Vapor-space ignitions hazards
		Corporate recommended/approved fire protection systems in place
		Flammable gases or liquefied flammable gasses
		Corporate recommended/approved fire protection systems in place
		Bulk dry chemicals
		Dust explosion potential addressed
		Tanks truck and railcar unloading and loading stations
		Spill containment and safe impounding
		Access platform safety
		Lighting adequate
		Grounding cables
		Fixed unloading pump and backflow preventer
		Emergency stop button location
		Connections lockable and closed

Column A Include	Column B Completed	Category/PSSR Item to Evaluate
		Placarding of hazards
		Remotely operated emergency stop valve for vehicles carrying hazardous materials
		Fusible-link fire valve on vehicles with bottom unloading of flammable
		Portable fire extinguisher at ground level or flammable
		Safety shower and eyewash units
		Recommended fire protection systems in place
		Electrical
		Process
		General workplace
		Safe operator access
		Building exits marked
		Lighting adequate
		Safety shower and eyewash units
		Accessible
		Located on each deck
		Located in control room
		Portable fire extinguishers
		Accessible
		Located on each deck
		Located in control room
		Human Factors
		Labeling of equipment, piping, critical valves, field instruments, switches
		Location of field instruments
		Sampling points
		Operator task safety
		Operator task ergonomics
		Opportunities for operator error
		Non-routine tasks
		Chemical Exposure Hazards
		Potential exposures
		Engineering controls adequate
		Building ventilation/fresh air intakes

APPENDIX A 127

Column A Include	Column B Completed	Category/PSSR Item to Evaluate
		Toxic gas monitors, alarms
		Protective equipment location
		Placarding
		Process Piping
		Construction appropriate for duty
		Materials quality assurance (including flange bolts), if critical, during construction
		Workmanship (for example, no short flange bolts)
		Routing satisfactory
		Adequately supported and guided
		Allowance for thermal expansion/no references
		No small diameter connections vulnerable to breakage/failure
		Expansion bellows properly installed/piping not able to move sideways/bellowed
		Undamaged during installation
		Flexible piping connectors correctly installed/undamaged (for example, kinked) during installation
		Necessary drains provided
		Hazardous outlets plugged closed
		Thermal (hydrostatic) pressure relief (including heat-traced sections)
		Sight glasses and gauge glasses
		External corrosion protection
		Freeze protection
		Insulation adequate for personal protection
		Protective flange covers
		Approved hoses and hose and connectors (no improvisations)
		Process Vents
		Flammable liquid breather vents provided with flame arrestors or conservation vents
		Telltale pressure gauge or other indicator provided between rupture disc and relief valve where a disc is installed below a relief valve
		Discharge piping from emergency pressure relief devices unrestricted by 90 degree ells,
		Excessive length or flame arrestors

Column A Include	Column B Completed	Category/PSSR Item to Evaluate
		Provisions such as drain holes to prevent accumulation of rainwater in discharge piping
		Discharge piping adequately supported to withstand reactive forces of pressure venting
		Safe vent discharge locations
		Manifolding of vents reviewed
		Vent sizing basis; documentation
		Ductwork
		Cleanouts
		Heat Exchangers, Jackets
		Vent, drains
		Thermal (hydrostatic) pressure relief
		Maintenance access (tube bundle)
		Machinery
		Guarding
		Local emergency stop button
		Emergency lubrication of critical machinery
		Maintenance provisions
		Local exhaust ventilation required for shaft seals
		Pumps
		Backflow prevention
		Connecting piping adequately supported to limit forces on casings
		Seal spray protection
		Isolation for maintenance
		Preparation for maintenance (drain and vent provided)
		Containment
		Spill containment
		Fire water runoff
		Process Controls/Control Room
		Control room inherent safety (vs. process hazards)
		Ventilation
		Emergency lighting
		Fire protection
		Field wiring security

APPENDIX A

Column A Include	Column B Completed	Category/PSSR Item to Evaluate
		System cable security
		Power supply
		Operator interface(s)
		Alarm systems
		Emergency shutdown
		Communications - normal and emergency
		Software access/security
		Software back-up
		Utilities Water Supply
		No municipal /potable water connections to the process
		Steam Boilers and Distribution
		Feedwater treatment chemicals handling
		Gas piping routing
		Combustion controls
		High and low drum water level alarms provided
		Bypass around Feedwater regulator accessible from operating level and located where
		Drum level gauge glass can be seen
		Two independent low water level trips provided for unattended boilers
		Dual safety relief valves in service
		Relief discharge piping adequately supported and drained
		Non-return valve on steam outlet
		Distribution piping – see Process piping
		Condensate drainage adequate
		Compressed Air Systems
		Non-lubricated construction or non-flammable synthetic lubricants used for compressor
		Discharge pressures above 100 psig
		Electrical
		Transformer location
		Motor control center(s)
		Standby Emergency Utility Systems
		Review provisions to satisfy proceeds safety requirements
		Waste Handling/Treatment

Column A Include	Column B Completed	Category/PSSR Item to Evaluate
		Inspect new facilities in the same manner as process facilities
		Warehouse
		Flammable and combustible liquids
		Forklifts and Motorized Hand Trucks
		Traffic safety
		Non combustible fuel
		Recommended Fire-Protection Systems in Place
		Maintenance Area and Shop
		General
		Local exhaust ventilation provided for welding
		Locker Room and Lunch Room
		Adequate space
		Provisions to protect contamination of food by chemicals
		Process Safety General
		Employee Participation Statement
		Process Safety Information
		Review of highly hazardous chemicals (HHC) and MSDSs
		Block flow diagram
		Maximum inventories
		Operating limits
		Equipment Information
		P&IDs
		Process Hazard Analysis (PHA) report(s)
		All PHA action items completed
		Training plan
		Contractor work rules
		Pre-Startup Safety Review plan
		Mechanical Integrity plan
		List of critical equipment
		Testing program with schedule
		Hot Work Permit System
		Site Management of Change Procedure
		Incident Investigation Plan

APPENDIX A

Column A Include	Column B Completed	Category/PSSR Item to Evaluate
		Emergency Action Plan (EAP)
		Facility EAP written
		Are new chemical or process hazards or risks such that changes to existing EAP are required?
		Do new facilities create any new transportation emergency response needs and are such needs in place? (Chemtrec update)
		Audit Schedule
		Operating Instructions
		Operating instructions clear and easily understood
		Instructions adequate (complete)
		Material hazards adequately covered for raw materials, catalysts, intermediates,
		Products and by-products
		Health hazards and permissible exposure levels (PELs)
		Physical hazards
		Handling precautions and safe handling procedures including Personal
		Protective equipment (PPE) requirements
		Corrective respiratory protection specified
		Process hazards adequately described
		Thermal hazards
		Any other hazards
		Tabulation of process alarms, interlocks (both safety & non-safety) and trips included
		Alarm and trip settings given
		Specific instruction included, or reference made to separate instructions, for
		Unloading and loading of bulk materials
		Step-by-step process procedures provided for each operating phase including:
		Initial start-up
		Normal start-up
		Normal operation
		Normal shutdown
		Emergency operations including emergency shutdown
		Start-up following emergency shutdown

Column A Include	Column B Completed	Category/PSSR Item to Evaluate
		Start-up following a turnaround
		Non-routine procedures (for example. equipment clean-out, equipment preparation for maintenance)
		Auxiliary equipment operation
		Operating limits clearly defined in step-by-step procedures
		Control ranges/limits specified
		Consequences of deviations given
		Responses to deviations/abnormal conditions specified
		Safe hold points specified
		PPE caution statements incorporated in step-by step procedures
		Use of checklists as appropriate
		Up-to-date
		All pages show revision number and date
		Reviewed for correctness
		Approved / signed by Department Manager
		Responsible Care
		Community Awareness and Emergency Response
		Communications training for key employees
		Education of employees on EAP, safety, health, and environmental
		Education of community on new process or change
		Outreach to educate responders, government officials, EAP
		Assessment of potential risks to employees from accidents
		Communication of emergency planning information to LEPC
		Facility tours for emergency responders
		Process Safety
		Current, complete documentation of process design and operating parameters

APPENDIX A

Column A Include	Column B Completed	Category/PSSR Item to Evaluate
		Current, complete documentation of hazards of materials and process
		Use of site management of change procedure
		Use of site incident investigation procedure
		Documented sound engineering practices consistent with recognized codes and standards
		Mechanical integrity program implemented for new unit or process change
		Employee Health and Safety
		Medical surveillance program tailored to meet needs of new process or change
		Personnel change to Central Safety Committee needed
		Pollution Prevention
		A quantitative inventory of wastes generated and releases to air, water and land
		Education of employees and public about the inventory and impact evaluation
		Documentation that waste generation is not increased by, or is minimized in, the change or new process
		Documentation that waste and release prevention objectives were included the design of the new modified process & products
		Distribution
		Review and training of distribution hazards with distributors
		Review of transportation routing to minimized potential risks
		Review with corporate transportation department
		Industrial Hygiene
		New substances
		Health care
		Toxicity data available
		Accident treatment plan
		Need for change in periodic medical exam

Column A Include	Column B Completed	Category/PSSR Item to Evaluate
		Occupational hygiene
		Inventory of possible sources of exposure
		Inventory means to restrict exposure
		Methods available to monitor exposure
		Suitable personal protection equipment available
		Hearing Conservation
		Noise level monitoring needed/arranged for new operations
		Engineering and administrative controls adequate
		Permissible exposure limits for chemical substances
		Appropriate exposure monitoring and evaluation arranged to determine compliance with applicable PELs
		PELs for mixtures applied when two or more hazardous substances present
		Engineering and administrative controls adequate
		Local exhaust ventilation systems
		Performance of local exhaust ventilation systems
		Local exhaust ventilation systems placed on inspection and maintenance program
		Control of chemical substances posing a potential occupational mutagenic or carcinogenic risk
		Are materials used having control levels A, B, C, or D?
		In the plant
		In the laboratory
		Hazard Communication Program
		Location inventory of chemicals updated
		Hazardous materials identified in accordance with definitions
		MSDSs on file and available to all employees
		Chemical containers labeled (or alternate means of label information provided)
		Piping labeled
		Training
		Respiratory protection

APPENDIX A 135

Column A Include	Column B Completed	Category/PSSR Item to Evaluate
		Review/confirm conformance with Corporate Industrial Hygiene Program
		Respirator selection in accordance with selection charts and specified in writing
		Pressure-demand SCBAs
		Emergency "escape only" respirators
		User medical clearance
		Facial hair policy
		Initial issues verified by supervision
		Fit testing
		Replacement of cartridges and canisters
		Inspection and maintenance
		Breathing air tested/tagged
		Training
		Smoking policy established
		Laboratory Control
		Process Laboratory Support Plan communicated
		Staffing adequate
		Laboratory facilities adequate
		General
		Emergency exits marked
		Emergency lighting
		Safety shower & eyewash
		Fire protection
		Laboratory equipment
		Suitable and adequate
		Maintenance provisions needed
		Storage and handling of chemicals
		Reagent storage
		Segregation adequate (oxidizers, acids)
		Flammable liquid storage
		Refrigerator for flammables explosion-proof
		Sample storage
		Sample and waste disposal
		Compressed gases

Column A Include	Column B Completed	Category/PSSR Item to Evaluate	
			Cylinder location safe (for example, away from heat sources)
			Quantities limited to immediate requirements
			Separation of flammable and oxidizers
			Toxic gas use limited to small cylinders
			Local exhaust ventilation for toxic gases
			SCBAs available for toxic gases
			Personnel trained in SCBA use as needed
		Laboratory procedures	
			Analytical procedures written and verified
			Sampling procedures included
			PPE requirements specified
			Training completed
			Industrial hygiene
		Maintenance	
			Necessary maintenance information in place
			Design drawings: for example as-built P&IDs, electrical schematics, isometric piping drawing
			Piping specifications
			Equipment purchase orders
			Equipment manuals
			Vendor prints
			Initial inspection and test results
			Resources adequate
			Needs communicated
			Plant personnel
			Contract maintenance (arrangements completed)
			Shop facilities
			Specialized requirements
			Requirements defined
			Skills available
			Equipment available
			Procedures developed
			Training completed

APPENDIX A

Column A Include	Column B Completed	Category/PSSR Item to Evaluate
		Certificate obtained and documented
		Maintenance management
		Service contracts arranged
		Maintenance management system in place
		Spare parts
		Requirements defined by maintenance department
		Procurement complete
		Start-up needs on hand
		Storage security
		Quality assurance program in place for critical equipment
		Materials of construction/quality of maintenance materials and parts
		Workmanship
		Preventive maintenance/mechanical integrity program developed
		Machinery
		Boilers and pressure vessels
		Critical equipment, vessels, piping, check valves, expansion bellows, flexible piping connectors, hoses defined
		Critical equipment inspection and test methods and frequencies defined
		Conservation vents, flame arrestors, PSEs, PSVs inspection and test methods and frequencies
		Critical instruments defined
		Proof-testing frequency
		Proof-testing procedures, validity
		Maintenance of combustion safety controls on direct-fired equipment
		Inspection and testing acceptance criteria developed and documented
		Training
		Operations and Maintenance
		Initial qualifications of personnel (knowledge and skills)
		Training program content vs. needs
		Safety orientation for new employees
		General safety training

Column A Include	Column B Completed	Category/PSSR Item to Evaluate
		Job-specific training
		Basic knowledge and skills
		Specialized knowledge and skills
		General process knowledge
		Material hazards, MSDSs
		Process hazards
		Process procedures
		Operating limits
		Consequences of deviations
		Responses to deviations/abnormal conditions
		Emergency procedures
		Field training
		Location of:
		Emergency equipment, showers, alarms
		Fire-fighting equipment
		Leak/spill prevention
		Reporting, mitigation
		Emergency drills
		Compliance with OSHA 1910.120 for hazardous waste operations
		New emergency response training requirements
		Measurement of training
		Effectiveness/certification (when applicable)
		Formal documentation of training
		Team assessment of training effectiveness
		Commissioning
		Commissioning plan and schedule
		Detailed, step-by-step plan written
		Plan adequately reviewed
		Responsibilities clearly defined and understood
		Plant verification of any equipment and systems check-out done by contractor
		Construction inspection by plant
		Confirm line-by-line conformity to P&IDs verified by plant, including:
		Materials of construction

APPENDIX A

Column A Include	Column B Completed	Category/PSSR Item to Evaluate
		Location of instrument elements/connections
		Orifice plate specifications and orientation
		Ranges of local PIs and TIs
		All local TIs have thermowells
		Identifies and relief pressure of PSEs & PSVs
		Actuated valve failure modes
		Equipment internals
		Vessels and piping
		Stress relieving done and documented as required
		Field radiography done and documented as required
		Pressure/Leak testing done and documented
		Cleaned and flushed (instruments, control valves, check valves protected)
		Special commissioning requirements (such as chemical cleaning, passivating, or testing)
		Vents and pressure relief valves
		Shipping supports removed from conservation vents
		Relief pressure of PSVs verified by test
		Fire protection systems
		Fire water systems inspection and commissioned
		Fire water pump acceptance test(s) completed and witnessed
		Other non-water fire protection system acceptance test(s) completed and witnessed (for example, CO2 or dry chemical
		New fire protection signaling systems and alarms commissioned
		Copies of completed test forms forwarded to insurance carrier and Corporate Risk Department
		New fire protection equipment, signaling systems and alarms placed on regular inspection and testing programs
		Electrical grounding
		Resistance of building and equipment and grounding systems measured <5 ohms
		Resistance to ground of all piping sections carrying flammable liquids and combustible powders measured <5 ohms
		Ventilation systems

Column A Include	Column B Completed	Category/PSSR Item to Evaluate
		Ventilation systems balanced
		Performance of local exhaust ventilation systems checked for conformance
		Machinery
		Alignment checked
		Absence of excessive forces on pump casings and other equipment from connected piping
		Pre-startup screens installed in pump suctions
		Agitator impeller security
		Lubrication systems functional
		Cooling systems functional
		Seal flush systems functional
		Rotation checked
		Vibration measurements
		Performance tests
		Other baseline data collection
		Instruments and control systems
		Program software checked
		Instrument loop sheet index available
		Pneumatic lines blown clean
		Loop checking done and documented
		Instrument calibrations done and documented review methods
		Alarm and trip points set and documented
		Interlocks (both safety & non-safety) tested
		Digital control system review and tests
		Combustion safety controls on direct-fired equipment
		Equipment inspection, adjustment and testing documented
		Punch list
		Review status
		Daily update
		Priorities with respect to start-up acceptable
		Water batching
		Plan developed

APPENDIX A

Column A Include	Column B Completed	Category/PSSR Item to Evaluate
		Start-up
		Start-up plan and schedule
		Written and reviewed
		Procedures for initial start-up specifically covered in the operating
		Instruction manual or under separate cover
		Reviewed and approved if separate
		Raw materials supply
		Technical support
		Adequate
		Lines and limits of authority clear
		Maintenance support
		Industrial hygiene monitoring
		Equipment monitoring
		Performance measurements and tests
		Regulatory Compliance
		New Substances
		Review of toxicity to environment
		Persistency in the environment
		Prevention of exposure to environment
		Destruction of substance when necessary
		Toxic Substances Control Act (TSCA)
		Project reviewed for any new requirements which might be imposed on the plant location under TSCA
		Transportation
		New transport operations adequately reviewed for compliance with all applicable DOT (or equivalent) regulations
		Emissions
		Operating permits obtained as directed by the site or corporate environmental departments
		Effluents
		New operations covered within the present NPDES permit or a new permit has been obtained
		Hazardous wastes

142 GUIDELINES FOR PERFORMING EFFECTIVE PRE-STARTUP SAFETY REVIEWS

Column A Include	Column B Completed	Category/PSSR Item to Evaluate
		Determined whether particular wastes qualify as hazardous wastes under federal, state and/or local laws and regulations
		On-site storage, treatment and/or disposal of hazardous wastes
		Reviewed for compliance with applicable laws and regulations, documentation?
		Off-site transportation and disposal of hazardous wastes reviewed with waste coordinator, documentation?
		PCBs
		Review equipment for PCB hazardous properties
		Supplementary Checklist for New Plant Sites
		Plant Security
		Access
		Fencing
		Visitor controls
		Vehicle controls
		Restrictions posted
		Communications
		Normal
		Emergency back-up
		Safety program
		Accountability
		Program conformance with Corporate Safety Standards
		Conformance with group safety standards
		First aid and emergency medical response
		Location procedures written
		Training conducted
		Industrial hygiene program
		Accountability
		Program conformance with Corporate Industrial Hygiene Standards
		Fire protection organization

APPENDIX A 143

Column A Include	Column B Completed	Category/PSSR Item to Evaluate
		Location organization
		Level of protection established in accordance with insurance requirement and Corporate Risk Department
		Training conducted in accordance with standard
		Equipment provided in accordance with standard
		Municipal fire department (or equivalent)
		Response time and capabilities consistent with location needs and fire protection organization
		Liaison established
		Familiarization
		Drills
		Process Safety Management
		Location coordinator appointed
		Program
		Training conducted

ATTACHMENT B - PSSR Potential Issue - Finding Form

Area under review:
Date of Review:
List of Sub-Team Members:
Instructions for using this form: For each issue, complete the information below. Electronically copy the blank fields as needed for each issue identified in Attachment A - *Process PSSR checklist*.

Description of potential issue or area of concern
Additional information (for example, the issue's criticality or a recommended solution)

PSSR Checklist Example A-4

DESIGN SAFETY REVIEW CHECKLIST
Dept.:
MOC ID#:
DATE:
Project ID:
INSTRUCTIONS: Check each question on the Yes or No line, or mark it N/A if not applicable. If an entire section of the checklist is not applicable, mark that section as N/A and no questions in that section need to be answered. No answers are considered deficiencies, and must be reported in writing to the Department Manager or designee and the change originator. This checklist is a guide to help identify possible deficiencies. All questions refer to the results, design and impact of the change, not broadly or in general to the system unaffected by the change. The reviewer is encouraged to look beyond the checklist for concerns which may be unique to the change and which may not be addressed here.

YES	NO	N/A	DSR CATEGORY SECTION/ITEM
			A. ADMINISTRATION
			1. Based on the current design, is the proposed change consistent with the original Process Hazard Analysis (PHA) assessment?
			2. Does the design comply with corporate standards?
			3. Has the impact of the change on existing buildings been considered? (That is, the design and location of new or modified equipment near occupied buildings, occupying a previously unoccupied building, and others.)
			4. Has any impact, beyond unit boundaries, associated with this change been properly dealt with and/or communicated?
			5. Have exposures to existing buildings (including pipe racks and cable trays) been considered when siting new vessels, utilities, temporary/permanent buildings or sheds, and others?
			6. Have noncombustible materials or construction been used?
			B. MATERIAL SAFETY/REGULATORY STATUS
			Have the following change scenarios been considered for possible Toxic Substance Control Act (TSCA) applicability?
			a. Previously non-isolated intermediates being temporarily isolated and/or held even for a short time, in non-hard-piped process equipment or in storage vessels (for example, drums).
			b. Previously non-isolated intermediates being held for an extensive period (for example, 24 hours or longer) in hard piped process equipment.
			c. Material previously burned or disposed of as a waste is reprocessed or sold.
			d. Different reactants or catalysts or different feed ratios are being used, thus producing different reaction products for TSCA purposes.

APPENDIX A 145

YES	NO	N/A	DSR CATEGORY SECTION/ITEM
			e. Change in the components or reactants in a polymer from < 2 % to > 2 % of the dry weight of the polymer produced.
			f. The TSCA Inventory status of different catalysts or reactants is unknown. If any of the above change scenarios are about to happen, contact the plant TSCA Coordinator immediately for an in-depth station of TSCA issues. Provide documentation of any concerns and their resolution.
			2. Have Material Safety Data Sheets (MSDS) or Preliminary Product Safety Data Sheets (PPSDS) been obtained for all chemicals to be handled, including isolated intermediates? (Consider changes in minor components of products and by-products)
			3. Has the potential for a hazardous chemical reaction in sumps and sewers been considered?
			4. Have all other potential product regulatory issues been addressed, for example, Department of Transportation (DOT), Federal Insecticide, Fungicide, and Rodenticide Act (FIFRA), Bureau of Alcohol, Tobacco, and Firearms (BATF), Food and Drug Administration (FDA) and ISO 9001?
			C. PRESSURE/VACUUM RELIEF
			(No relief devices in this project)
			1. Have new or modified safety relief device(s) or vent system(s) been designed in accordance with Plant Engineering and Site requirements?
			2. Has potential for external pressure (vacuum) from sudden cooling, condensing, pump-out, during clean-up or preparation of equipment for maintenance, or potentially submerged overflow line been addressed?
			3. Have only full-port valves been specified for use at the inlet and outlet of pressure/vacuum relief devices?
			4. Have any changes to safety relief device inlet or outlet piping been properly reviewed?
			5. Will adequate facilities (alarms, detectors, redundancy, and others.) be provided to minimize the risk of a relief device actuating due to equipment, instrumentation, or utility failure?
			6. Have the discharges of safety relief devices been located so as to avoid potential personnel injury and damage to associated equipment?
			7. Has the design included installation of the safety relief valve vertically?
			D. TEMPERATURE/REACTION
			1. Has potential for formation of unwanted by-products been adequately addressed?
			2. Has potential for loss of flow or reverse flow been adequately addressed?

YES	NO	N/A	DSR CATEGORY SECTION/ITEM
			3. Have adequate provisions been made so that normally dilute but reactive materials CANNOT be concentrated or accumulated in unexpected areas (stagnant pipe/valves; utility systems; feed or reaction vessels, sewers, and others)?
			4. Is adequate freeze protection provided?
			E. VALVES AND PIPING
			1. Have the proper valve and piping specifications been used?
			2. Have cross-tied lines (pump headers, utility lines, between high/low pressures, and others.) been reviewed to minimize contamination potential and eliminate mixing of reactive chemicals?
			3. Have test methods and documentation requirements been specified to ensure the integrity of new and revised piping systems?
			4. Will sample points be properly configured for safe sampling of hazardous chemicals?
			5. Have all open ended valves and hand-operated ball valves been designed in accordance with environmental requirements (that is, NESHAPS covered materials such as benzene and formaldehyde) or plant standards?
			6. Have hot-taps been reviewed and eliminated where possible?
			7. Will necessary excess flow and back-flow prevention measures be provided?
			8. Has line expansion and vibration during startup, shutdown, cleaning, normal operation, and others, been considered, and, if appropriate, analyzed in detail?
			9. Has the potential risk and consequences of hydraulic hammering been considered?
			10. Have appropriate materials of construction been considered for compatibility, corrosion resistance and GMP requirements? (Consider 0-rings, gaskets, diaphragms, and others.)
			11. Have the temporary start-up strainers been identified to ensure removal for normal operation?
			12. Has the design included the proper encasing of underground piping at stress points (for example, roadways, railroads, and others.)?
			F. ROTATING AND MECHANICAL EQUIPMENT
			1. Have special precautions for safe operation of equipment been considered? (Reverse flow, minimum flow, maximum head, sudden flow increase during startup, total recycle, vapor locking, volute drainage, Emergency Shutdown Device(s) (ESD), and others.)
			2. Do new and revised pumps and/or pump seals meet and GMP requirements and corporate standards?

APPENDIX A

YES	NO	N/A	DSR CATEGORY SECTION/ITEM
			3. Have lubricants and buffer fluids been properly selected to meet any GMP requirements (that is, food grade) such that leakage into the process will not result in undesired chemical reactions or product contamination?
			4. Will moving parts on machinery be properly guarded?
			5. Will pumps be located in accordance with corporate standards?
			6. Are the emergency shutdown systems (over speed, ground fault, high temperature, vibration, and others.) adequate?
			7. Have appropriate materials of construction been considered for compatibility, corrosion resistance and GMP requirements? (Consider 0-rings, gaskets, diaphragms, and others.)
			8. Will adequate pressure relief be provided for new or modified pump systems?
			G. INSTRUMENTATION
			1. Has the potential for instrument failure been adequately addressed?
			2. Have all new shutdown devices been designed to permit testing.
			3. Have potential consequences of instrument or computer failure (redundancy, backup power supply, or others) been considered?
			4. Has control valve fail-safe position on loss of electric power or air/nitrogen been properly specified to minimize the impact of failure?
			5. Have provisions been made to safeguard against risks associated with control valves going full open or closed (mechanical stops, limit switches, and others.)?
			6. If the change affects a shutdown or ESD system, have issues been addressed?
			7. Will the alarms associated with critical instruments be clearly displayed in the control room?
			8. Have any special concerns with regard to response time or sequencing been adequately addressed? (Consider over speed trips, hydraulic hammer, surge control, and others.)
			9. Have process changes (capacity change, density, viscosity, vapor pressure, and others) been considered in the design of new and existing instrumentation?
			10. Has ESD control logic been reviewed?
			11. Will temperature elements be mounted in thermo wells?
			12. Have appropriate materials of construction been considered for compatibility, corrosion resistance and any GMP requirements? (Consider 0-rings, gaskets, diaphragms, and others.)
			13. Has adequate local instrumentation been addressed for safety and trouble-shooting purposes?
			14. Is the location of sensing elements proper to ensure that they are actually measuring what you want measured?

YES	NO	N/A	DSR CATEGORY SECTION/ITEM
			H. ELECTRICAL SYSTEMS
			1. Have instrumentation and electrical equipment enclosures been specified to meet the electrical classification of the area?
			2. Have wire gauges, starters and overloads been properly sized in accordance with the National Electric Code (NEC)?
			3. Has the design considered the requirements of electrical hot work and cranes near power lines such that outage requirements and variances will be minimized?
			4. Has the design included adequate room for ventilation of transformers, motors, and other similar equipment?
			I. FIRE PROTECTION
			1. Has the potential for static electricity buildup been adequately addressed?
			2. Has proper grounding of all electrical and process equipment (including piping and shipping containers) been specified?
			3. Will fire containment be adequate where hazardous or reactive chemicals are present which can result in high energy release when mixed, contaminated, heated or otherwise mishandled?
			4. Has spontaneous heating of leakage into insulation been adequately considered?
			5. Are provisions made for safe handling of flammable or potentially explosive materials? Consider materials of construction, cleanup and preparation of equipment for maintenance.
			6. Are all fire water spray-system modifications being designed (including flow calculations) and installed by qualified personnel or contractor?
			7. Have modifications or additions to fire protection systems been reviewed or accepted by the safety department or property insurance carrier?
			8. Will vents potentially containing flammables be provided with adequate safety equipment (flame arrestors or inert gas purge system)?
			9. Will an adequate detection (vapor or explosive gas detectors) and response system be provided where a vapor cloud is likely?
			10. Have the risks associated with any ignition source or explosive gas mixture been adequately dealt with?
			11. Has adequate fire safety equipment been specified and located where needed?
			12. Has deluge water overflow from containment systems been determined and reviewed properly?
			13. Is diking, curbing, or drainage adequate to contain spills and contaminated rainwater? [Resource Conservation and Recovery Act (RCRA)]

APPENDIX A

YES	NO	N/A	DSR CATEGORY SECTION/ITEM
			J. PERSONNEL HEALTH & INDUSTRIAL HYGIENE
			1. Does the design adequately consider medical, industrial hygiene, ergonomic and GMP
			factors (heat stress, biological stress, high noise level, poor lighting, adequate ventilation, potential of oxygen deficient atmosphere, difficult tasks, repetitive motion tasks, poor equipment access or egress, exposure to hot surface [> 140?F], dust hazards, and others.)?
			2. Have adequate provisions been specified for the safe handling and sampling of corrosive, toxic, carcinogenic, teratogenic, or otherwise hazardous materials?
			3. Have personnel safety devices (showers, eye baths, fall prevention, breathing air systems) been specified?
			4. Will all asbestos-containing insulation be removed/repaired in accordance with plant requirements?
			5. Will secondary surfaces (grating, slip protection coatings) be provided where freezing or slippery materials are handled?
			6. Has confined space entry (including electrical equipment) been adequately addressed in this design?
			7. Has the design included adequate walking/working surfaces in accordance with OSHA standards and plant requirements?
			8. Is potable water kept physically separated from process usage?
			K. WASTE WATER TREATMENT & SPILL PREVENTION
			1. Does this design address the potential for spills, releases, compatibility and flammability?
			2. Has the Environmental Health and Safety Department (EHS) been notified of any new waste stream sources or increased quantities from existing sources?
			3. Has EHS evaluated waste streams for impact on Plant Waste Water Treatment Facilities? (Compatibility with existing plant waste treatment system capabilities, required provisions for RCRA hazardous waste chemicals, proper notification of governmental agencies, effect on EHS operations personnel, effect on discharges to the Publicly Owned Treatment Work(s) (POTW) or river, and others.)
			4. Will diking, draining, curbing, and special protective surface coatings be adequate to contain leaks and the worst case spill scenario?
			L. SOLID & LIQUID WASTE
			1. Has at-source waste minimization been adequately addressed?
			2. Have adequate provisions been made for disposal of all wastes (including drums, bags, filter elements, liquid residues, asbestos-containing insulation, PCBs, contaminated soil, demolition rubble and processing equipment)?

YES	NO	N/A	DSR CATEGORY SECTION/ITEM
			3. Have waste streams intended for disposal in the boilers been reviewed by the EHS department?
			4. Have the following change scenarios been considered for possible RCRA applicability?
			a. Creation of a new hazardous waste.
			b. Change in composition of an existing hazardous waste.
			c. Modifications of a facility in hazardous or solid waste service. If any of the above change scenarios are about to happen, contact the plant EHS department immediately for an in-depth examination of RCRA issues. Provide documentation of any concerns and their resolution.
			M. AIR EMISSIONS
			1. Have the following change scenarios been considered for possible air emission applicability?
			a. Increase in potential emissions of volatile organic compound
			b. New emissions of hazardous air pollutants
			c. Increase in potential emissions of hazardous air pollutants
			d. Change of service for equipment
			e. Composition change in existing emissions
			f. Addition of new emission points (including fugitive sources such as valves and relief devices) or physical change in existing emission points or monitored points. If any of the above change scenarios are about to happen, contact the plant EHS department immediately for an in-depth examination of permit issues. Provide documentation of any concerns and their resolution.
			2. Have any increases in air emissions of flammable, toxic, corrosive, reactive or otherwise hazardous chemicals been identified, quantified and communicated to the plant EHS department (to obtain necessary permitting, identify control devices being used, and others)?
			3. Has the change been reviewed to determine if it is subject to any requirements to recalculate Toxic Release Inventory (TRI), re-evaluate control device efficiency, or otherwise manage the change under state or federal requirements?
			N. PROCESS EQUIPMENT
			1. Has a pressure vessel engineer reviewed the design/repair specifications and considerations for new, altered or repaired equipment?
			2. Have the appropriate welding procedure(s) been identified?
			3. Has spare equipment been provided where needed for safety?
			4. Have appropriate materials of construction been considered for compatibility, corrosion resistance and any GMP requirements? (Consider 0-rings, gaskets, diaphragms, and others.)

APPENDIX A 151

YES	NO	N/A	DSR CATEGORY SECTION/ITEM
			5. Has the documentation for piping and equipment been appropriately updated?
			O. COMPUTER SOFTWARE AND SYSTEMS
			1. Are adequate safeguards in place to ensure the process is controlled within the safe operating envelope?
			2. Is the fail-safe condition of controllers adequate?
			P. OTHER SAFETY CONSIDERATIONS
			1. Will adequate design provisions exist for cleanup and preparation of equipment for maintenance of equipment/piping/control systems (including lockout of all energy sources and double isolation where required)?
			2. Have the consequences of loss of any utility been adequately addressed?
			3. Has cathodic protection, electrical continuity and electronic grounding been adequately addressed?
			4. Has necessary structural analysis been performed?
			5. Will adequate provisions be made to ensure equipment idled by this change is maintained in a safe condition?
			6. Has a change in utility consumption been properly communicated?
			7. Has the design included the need for labeling including equipment apparatus numbers, identity, content, direction of flow, and others?
			8. Has the design included requirements for adequate area lighting?
			9. Has the design of the equipment included adequate room and clearance for proper maintenance?
			Q. FINAL QUESTIONS
			1. Have these guide questions adequately addressed all areas of concern?
			2. Other:
			3. Other:

DESIGN SAFETY REVIEW (DSR) SUMMARY
Question # (add as needed)
Explanation
Additional Comments:
Date:
Design Safety Reviewer:
Date:
First-Level Manager:

DSR SIGNATURES
The design safety review is complete and the change is recommended for implementation.
Date of Review:
DSR Chairperson: (print name and sign)
Team Members: (print name, sign, and list team function)

APPENDIX B – INDUSTRY REFERENCES

American Chemical Society
www.acs.org
1155 16th Street NW
Washington DC 20036
(202) 872-4600
(800) 227-5558

American Industrial Hygiene Association
www.aiha.org
2700 Prosperity Ave., Suite 250
Fairfax, VA 22031
(703) 849-8888

American Institute of Chemical Engineers
Center for Chemical Process Safety
www.aiche.org
3 Park Ave.
New York, N.Y., 10016-5991
(800) 242-4363

American Petroleum Institute
www.api.org
1220 L Street NW
Washington, DC 20005-4070
(202) 682-8000

American Chemistry Council
www.americanchemistry.com
1300 Wilson Boulevard
Arlington, VA 22209
(703) 741-5000

National Safety Council
www.nsc.org
1121 Spring Lake Drive
Itasca, IL 60143-3201
(630) 285-1121

Synthetic Organic Chemical Manufacturers Association (SOCMA)
www.socma.com
Synthetic Organic Chemical Manufacturers Association
1850 M St N.W., Suite 700
Washington, D.C. 20036
(202) 721-4100

United States Chemical Safety and Hazards Investigation Board
www.csb.gov
2175 K Street N.W., Suite 400
Washington, D.C. 20037-1809
(202) 261-7600

United States Department of Labor, Occupational Safety and Health Administration
www.osha.gov
Washington, DC 20210
(800) 488-7087

United States Department of Transportation
www.dot.gov
400 Seventh Street, SW
Washington, DC 20590
(202) 366-4000

APPENDIX B

United States Environmental Protection Agency
www.epa.gov
401 M St. S.W.
Washington, DC 20460
(202) 260-2090

APPENDIX C

APPENDIX C – REGULATORY REFERENCES

- Environmental Protection Agency, *Accidental Release Prevention Requirements: Risk Management Programs*, 40 CFR 68, Clean Air Act, Section 112 (r)(7), Washington, DC, 1996.

- Occupational Safety and Health Administration, *Process Safety Management of Highly Hazardous Chemicals*, 29 CFR Part 1910, Section 119, Washington, DC, 1992.

- OSHA Instruction CPL 2-2.45A CH-1 September 13, 1994 Directorate of Compliance Programs, Subject: 29 CFR 1910.119, *Process Safety Management of Highly Hazardous Chemicals – Compliance Guidelines and Enforcement Procedures.*

INDEX

A

Action Item Tracking Systems, 83-84
Action Items for Process Hazards Analysis, 77
Administrative Procedure for PSSR, 70-76
Algorithm for Risk-Based PSSR, 40-41, 52, 54
Approval, Procedures for, 79-80
Associated Risk, 1, 82
Audit Action Items, 104
Audit Protocols, 100-101
Auditing PSSR Systems, 92-93, 98-101

B

Benefits, 1-4, 9, 17-18, 21, 24, 28
Best Practices for PSSR, 27-28, 79, 94

C

Capital Projects, 18-19
Categorizing Changes Requiring PSSR, 26
Center for Chemical Process Safety (CCPS), 6, 23
Changes,
  Definition of, 24
  Risk Evaluation of, 31-32
  Temporary, 20
Changes Requiring PSSR, 26
Changes to:
  Chemicals Used, 39
  Control Systems, 43
  Equipment and Instrumentation, 42-43
  Operating Facilities, 19
  Software, 42
Checklist Compilation Methodologies, 82-91

Checklist Templates, 106-153
Chemical Considerations, 37, 39
Codes and Standards, 2, 5-6, 24-26, 28, 34, 100-101, 103
Communicating Problems and Solutions, 93-94
Complex/Long Form Reviews, 13, 33, 46-51
Complexity, Affects on Team Size and Expertise, 37-39
Compliance Documentation, 13
Consequence Analyses, 16, 78
Continuous Improvement, 92-104
Critical Items for Safe Operation, 79

D

Databases for
  Documentation, 28, 34, 42
  PSSR Questions, 82-83
  Review Items, 83
Definition of
  PSSR System, 64-65
  PSSR Policy, 68
Description of PSSR Team, 71
Design of Specific PSSR, 69-70
Diagnosis of PSSR System Issues, 92-93
Documentation, 14, 25, 27, 34, 42-43, 52-53, 64-65, 67, 76, 78

E

Efficiency Improvements, 94
Efficiency Metrics, 97-98
Electronic Change Management Systems, 84-85
Electronic Checklist Tools, 84-91
Emergency Procedures, 7, 27, 67
Emergency Shutdowns, 21-22,
Environmental Considerations, 19, 25, 44
EPA RMP Regulations, 11, 21, 23-24, 40, 41, 43, 65-67, 78, 95-100, 104
Equipment
  Modification, 16, 54
  Risk Considerations of, 37

H

Hardware Changes or Trigger Events, 1, 42
Hazardous Materials Storage, 29
Health Considerations, 3, 6, 16, 29, 44, 56
Higher Risk Trigger Events, 54

I

Impact of Changes to Operating Processes, 19
Incident Consequences, 16
Industry References, 154-156
Interface with PSM Elements, 4-5

L

Lay-up Procedures, 20
Long Form Reviews, 13, 16, 33, 39, 41, 44, 46-52, 55, 69-70, 78-79

M

Maintenance
 Activities, 21, 39
 Procedures, 27, 39
Management of Change, 27, 37, 40, 52, 55-56, 65-67, 80, 84, 98
Management Team Training, 17-18
Matrix System, Priority Risk-Ranking, 33
Mechanical Integrity, 4, 65
Modifying PSSR Processes, Reasons for, 94-96
Mothballed Processes, Restarting of, 20

N

Novelty, 33, 36-37

O

Occupational Health Considerations, 3, 6, 16, 29, 56
Operating Procedures, 21, 27, 39
OSHA Process Safety Management Regulations, 4, 6-7, 21-24, 40, 64, 66-68, 97, 101

P

Performance Based Regulations, 1, 31
Performance Metrics, 97-98
Personal Protective Equipment, 29, 39
Policy Statements, 68
Post-Review Action Item Follow-up, 41
Priority Risk-Ranking Matrix System, 3
Process Algorithm, 40-41

Process Equipment, Mothballing of, 20
Process Hazards Analysis, 27, 40, 67, 77-78
Process Modification, Sample PSSR of, 54-61
Process Safety Information, 24, 26, 42-43, 55
PSM Compliance Considerations, 27
PSSR Checklists, 85-91
PSSR Related Documents, 25, 76
PSSR Roles and Responsibilities, 66, 71
PSSR Scheduling Considerations for
  Capital Projects, 18
  Changes to Operating Facilities, 19
  Post-turnaround Startup, 21
  Restarting Mothballed Processes, 20
  Routine Maintenance, 21
  Startup after Emergency Shutdown, 21
  Temporary Changes, 20

Q

Qualitative Approach to Risk-Based PSSR, 39
Qualitative Risk Assessment, 31-32, 36
Qualitative Risk Levels, 39
Qualitative Tools, 6, 31
Quality Assurance, 38
Quantitative Risk Assessment, 3, 32
Questionnaires, Risk-Based Ranking, 33, 44

R

RAGAGEP, 2
Redundant Verification, 22
Regulation
 EPA, 6, 23-24, 64, 66, 68
 International, 23
 OSHA, 4, 6-7, 21-24, 40, 64, 66-68, 97, 101
Regulatory References, 157
Replacement in Kind, 12, 24,
Review Level and Scope, 39
Risk
  Analysis of, Qualitative, 31-32
  Analysis of, Quantitative, 32
  Analysis of, Semi-Quantitative, 32
  Analysis Techniques, 31-63
  Management Program Regulation (EPA), 21, 23-24, 40, 64, 78
  Ranking Questionnaires, 33, 44
Risk-Based

INDEX **163**

Approaches to PSSR, 6, 16, 31-63
  Decision Guidelines, 39
  Decision Making Steps, 34
  Ranking, 32-33, 42
RMP Compliance Considerations, 27
Roles and Responsibilities for PSSR, 66
Routine Maintenance Activities, 21

S

Safe Work Practices, 19, 38-40, 43, 53, 76
Safety Considerations, 29, 44, 67
Safety Systems, Impact of, 26, 69
Scheduling Considerations, 18-22, 32, 69
Scope of Review, 39
Security Considerations, 29-30
Security Vulnerability Analysis, 30
Selection Criteria for Review Type, 16
Simple/Short Form PSSR, 13, 33, 37, 41, 44-45, 51-53, 69-70, 78-79
Software Changes or Trigger Events, 42, 77
Software Tools, 84-85
Startup Procedures, 20-21

T

Team Leader Training, 17
Team Size and Expertise, 37, 68-69
Techniques for Risk Analysis, 31-63
Temporary Changes, 20
Tie-in Points, 16, 19
Tracking and Authorization Systems, 84-85
Training Documentation, 19, 70
Training for Continuous Improvement, 93-94
Training,
  Instructional Design for, 43, 70
  Learning Objectives for, 17
  Quality and Verification of, 43-44, 67
Training of
  Management Team, 17, 70
  Operating Teams, 27, 67
  PSSR Team Leaders, 17, 70
  Team Leaders and Members, 17, 19
  Workforce, 67, 70
Trigger Events, 12-13, 29, 31-32, 36-41, 44, 52, 65, 68, 78
Turnaround Procedures, 21

U

Upgrading PSSR Systems, Suggestions for, 96

V

Verification of Completion, 77

W

Waste Characterization, 29
Work Process for PSSR, 15, 64-80

CUSTOMER NOTE: IF THIS BOOK IS ACCOMPANIED BY SOFTWARE, PLEASE READ THE FOLLOWING BEFORE OPENING THE PACKAGE.

This software contains files to help you utilize the models described in the accompanying book. By opening the package, you are agreeing to be bound by the following agreement:

This software product is protected by copyright and all rights are reserved by the author and John Wiley & Sons, Inc. You are licensed to use this software on a single computer. Copying the software to another medium or format for use on a single computer does not violate the U.S. Copyright Law. Copying the software for any other purpose is a violation of the U.S. Copyright Law.

This software product is sold as is without warranty of any kind, either express or implied, including but not limited to the implied warranty of merchantability and fitness for a particular purpose. Neither Wiley nor its dealers or distributors assumes any liability of any alleged or actual damages arising from the use of or the inability to use this software. (Some states do not allow the exclusion of implied warranties, so the exclusion may not apply to you.)

WILEY